Technical Note 1431

A Technical Reference for CFAST: An Engineering Tool for Estimating Fire and Smoke Transport

Walter W. Jones
Glenn P. Forney
Richard D. Peacock
Paul A. Reneke

National Institute of Standards and Technology
Building and Fire Research Laboratory
Gaithersburg, MD 20899

April 2003

U.S. Department of Commerce
William M. Daley, *Secretary*
Technology Administration
Dr. Cheryl L. Shavers, *Under Secretary for Technology*
National Institute of Standards and Technology
Raymond G. Kammer, *Director*

Bibliographic Information

Abstract

CFAST is a zone model capable of predicting the environment in a multi-compartment structure subjected to a fire. It calculates the time evolving distribution of smoke and fire gases and the temperature throughout a building during a user-specified fire. This report describes the equations which constitute the model, the physical basis for these equations, data which are used by the model, and details of the operation of the computer program implementing the model. The means by which one can add new phenomena are detailed, as are the variables and structure of the model.

A set of comparisons between the model and a range of real-scale fire experiments is presented. In general, the CFAST model compares favorably with the experiments examined in this report. Although differences between the model and the experiments were clear, they can be explained by limitations of the model and of the experiments.

Keywords

Computer models; computer programs; evacuation; fire models; fire research; hazard assessment; human behavior; toxicity

Ordering Information

National Institute of Standards and Technology
Technical Note 1431
Natl. Inst. Stand. Technol.
Tech. Note 1431
178 pages (April 2003)

U.S. Government Printing Office
Washington: 2003

For sale by the Superintendent of Documents
U.S. Government Printing Office
Washington, DC 20402

DISCLAIMER

The U. S. Department of Commerce makes no warranty, expressed or implied, to users of CFAST and associated computer programs, and accepts no responsibility for its use. Users of CFAST assume sole responsibility under Federal law for determining the appropriateness of its use in any particular application; for any conclusions drawn from the results of its use; and for any actions taken or not taken as a result of analyzes performed using these tools.

Users are warned that CFAST is intended for use only by those competent in the field of fire safety and is intended only to supplement the informed judgment of the qualified user. The software package is a computer model which may or may not have predictive value when applied to a specific set of factual circumstances. Lack of accurate predictions by the model could lead to erroneous conclusions with regard to fire safety. All results should be evaluated by an informed user.

INTENT AND USE

The algorithms, procedures, and computer programs described in this report constitute a methodology for predicting some of the consequences resulting from a specified fire. They have been compiled from the best knowledge and understanding currently available, but have important limitations that must be understood and considered by the user. The program is intended for use by persons competent in the field of fire safety and with some familiarity with personal computers. It is intended as an aid in the fire safety decision-making process.

CONTENTS

Figures and Tables

Nomenclature

α thermal diffusivity (m^2/s)

β expansion coefficient, $1/T_f (K^{-1})$

ν kinematic viscosity (m^2/s)

A_s surface area (m^2)

C   correlation coefficient

g   gravitational constant (9.8 m/s$^{2)}$

h   convective heat transfer coefficient (W/m^2 K)

k   equivalent thermal conductivity of air (W/m K)

L   characteristic length of the geometry (m)

Gr_L Grashof number (dimensionless)

Nu_L Nusselt number (dimensionless)

Pr   Prandtl number (dimensionless)

$\dot{q}$ convective heat transfer rate (W)

Ra_L Rayleigh number (dimensionless)

T_f film temperature (K)

T_g bulk gas temperature (K)

T_s surface temperature (K)

y_p pyrolysis front position in the wind aided direction.

x_p pyrolysis front position in the opposed flow direction.

y_b burnout front position in the wind aided direction.

x_b burnout front position in the opposed flow direction.

y_f flame tip position.

T_{layer} temperature of adjacent gas layer.

T_{ig} temperature at which the surface ignites.

T_s temperature of the surface.

$T_{s,min}$ minimum temperature at which lateral flame spread occurs.

Q''_{TOT} total heat per unit area

$\dot{Q}''$ heat Release Rate per unit area

$\dot{q}_f''$	flux per unit area from the flame
ΔL	effective heat of gasification
σ	Stefan-Boltzmann constant
k	conductivity
ρ	density
c	cpecific heat
Φ	Lateral flame spread parameter

A_d	duct surface area (m^2)
A_o	area of the inlet, outlet, duct, contraction, or expansion joint, coil, damper, bend, filter, and so on in a mechanical ventilation system. (m^2)
A_{room}	floor area of a room (m^2)
A_{slab}	cross-sectional area for horizontal flow (m^2)
A_v	area of ceiling or floor vent (m^2)
A_w	wall surface area (m^2)
b_i	coefficients for adsorption and desorption of HCl
C	flow coefficient for horizontal flow of gas through a vertical vent
C_{LOL}	Lower oxygen limit coefficient, the fractional burning rate constrained by available oxygen, eq (39)
C_o	characteristic flow coefficient
C_w	wind coefficient – dot product of the wind vector and vent direction
CO/CO_2	ratio of the mass of carbon monoxide to the mass of carbon dioxide in the pyrolysis of the fuel
CO_2/C	ratio of the mass of carbon dioxide to the mass of carbon in the pyrolysis of the fuel
c_k	heat sources for the k'th wall segment (W)
c_p	heat capacity of air at constant pressure $(J/kg\ K)$
c_v	heat capacity of air at constant volume $(J/kg\ K)$
D	effective diameter of ceiling or floor vent (m)
D_e	effective duct diameter (m)
d_{HCl}	rate of deposition of HCl onto a wall surface, eq (150) (kg/s)
E_i	internal energy in layer i (W)
F	friction factor
F_{k-j}	configuration factor
g	gravitational constant (m^3/s)

G	conductance
Gr	Grashof number
H_c	heat of combustion of the fuel (J/kg)
h_c	convective heat transfer coefficient (J/m^2 K)
$\hbar_i$	rate of addition of enthalpy into layer i (W)
h_l	convective heat transfer coefficient in ceiling boundary layer (J/m^2 K)
$\hbar$	characteristic convective heat transfer coefficient
H	height of the ceiling above a fire source (m)
H/C	ratio of the mass of hydrogen to the mass of carbon in the pyrolysis of the fuel
H_2O/H	ratio of the mass of water to the mass of hydrogen in the pyrolysis of the fuel
HCl/C	ratio of the mass of hydrogen chloride to the mass of carbon in the pyrolysis of the fuel
HCl/f	ratio of the mass of hydrogen chloride to the total mass of the fuel
HCN/C	ratio of the mass of hydrogen cyanide to the mass of carbon in the pyrolysis of the fuel
HCN/f	ratio of the mass of hydrogen cyanide to the total mass of the fuel
k	mass transfer coefficients for HCl deposition
l	characteristic length for convective heat transfer (m)
m_i	total mass in layer i (kg)
$m_{i,j}$	mass flow from node i to node j in a mechanical ventilation system (kg/s)
$\dot{m}_b$	burning rate of the fuel (perhaps constrained by available oxygen) (kg/s)
$\dot{m}_c$	production rate of carbon during combustion (kg/s)
m_d	mass flow in duct (kg/s)
$\dot{m}_e$	rate of entrainment of air into the fire plume (kg/s)
$\dot{m}_f$	pyrolysis rate of the fuel (before being constrained by available oxygen) (kg/s)
$\dot{m}_i$	rate of addition of mass into layer i (kg/s)
O/C	ratio of the mass of oxygen to the mass of carbon in the pyrolysis of the fuel
P	pressure (Pa)
P_{ref}	reference pressure (Pa)
Pr	Prandtl number
Q_c	total convective heat transfer (W)
Q_{eq}	dimensionless plume strength at layer interface
Q_f	total heat release rate of the fire (W)
Q_H	dimensionless plume strength at the ceiling
Q_r	total radiative heat transfer (W)
r	radial distance from point source fire (m)

R   universal gas constant (J/kg K)

R_e Reynolds number

S   vent shape factor for vertical flow

S/C   ratio of the mass of soot to the mass of carbon in the pyrolysis of the fuel

t   time (s)

T_{amb} ambient temperature (K)

T_d duct temperature (K)

T_e temperature of gas entrainment into the fire plume (K)

T_g gas temperature (K)

T_i temperature of layer i (K)

T_{in} duct inlet temperature (K)

T_k temperature of the k'th wall segment (K)

T_{out} duct outlet temperature (K)

T_p temperature of the plume as it intersects the upper layer (K)

T_w wall temperature (K)

v   gas velocity (m/s)

V_d duct volume (m³)

V_i volume of layer i (m³)

Y   mass fraction of a species in a layer

Y_{LOL} lower oxygen limit for oxygen constrained burning, expressed as a mass fraction

z   height over which entrainment takes place (m)

Z   height (m)

α absorption coefficient of the gas (m⁻¹)

ΔP pressure offset from reference pressure, P - P_{ref} (Pa)

γ ratio of c_p/c_v

ϵ_k emissivity of the k'th wall segment

κ thermal conductivity (J/m s K)

ν kinematic viscosity (m²/s)

ρ_d density of gas in a duct (kg/m³)

ρ_i density of gas in layer i (kg/m³)

σ Stefan-Boltzman constant (5.67 x 10⁻⁸ W/m²K⁴)

τ transmissivity factor

χ_c fraction of the heat release rate of the fire which goes into convection

χ_r fraction of the heat release rate of the fire which goes into radiation

A Technical Reference for CFAST: An Engineering Tool for Estimating Fire and Smoke Transport

Walter W. Jones, Glenn P. Forney, Richard. D. Peacock and Paul A. Reneke

Building and Fire Research Laboratory
National Institute of Standards and Technology

1 Overview

Analytical models for predicting fire behavior have been evolving since the 1960's. Over the past decade, the completeness of the models has grown considerably. In the beginning, the focus of these efforts was to describe in mathematical language the various phenomena which were observed in fire growth and spread. These separate representations have typically described only a small part of a fire. When combined though, they can create a complex computational model intended to give an estimate of the expected course of a fire based upon given input parameters. These analytical models have progressed to the point of providing predictions of fire behavior with an accuracy suitable for most engineering applications. In a recent international survey [1], 36 actively supported models were identified. Of these, 20 predict the fire driven environment (mainly temperature) and 19 predict smoke movement in some way. Six calculate fire growth rate, nine predict fire endurance, four address detector or sprinkler response, and two calculate evacuation times. The computer models now available vary considerably in scope, complexity, and purpose. Simple "compartment filling" models such as the Available Safe Egress Time (ASET) model [2] run quickly on almost any computer, and provide good estimates of a few parameters of interest for a fire in a single compartment. A special purpose model can provide a single function. For example, COMPF2 [3] calculates post-flashover compartment temperatures and LAVENT [4] includes the interaction of ceiling jets with fusible links in a compartment containing ceiling vents and draft curtains.

In addition to the single-compartment models mentioned above, there are a smaller number of multi-compartment models which have been developed. These include the BRI transport model [5], FAST [6], CCFM [7] and the CFAST model discussed below [8].

Although the papers are several years old, Mitler [9] and Jones [10] reviewed the underlying physics in several of the fire models in detail. The models fall into two categories: those that start with the principles of conservation of mass, momentum, and energy such as CFAST; and those that typically are curve fits to particular experiments or series of experiments, used in order to discern the underlying relationship among some parameters. In both cases, errors arise in those instances where a mathemati-

1

cal short cut was taken, a simplifying assumption was made, or something important was not well enough understood to include.

Once a mathematical representation of the underlying science has been developed, the conservation equations can be re-cast into predictive equations for temperature, smoke and gas concentration and other parameters of interest, and coded into a computer for solution.

The environment in a fire is constantly changing. Thus the equations are usually in the form of *differential equations*. A complete set of equations can compute the conditions produced by the fire at a given time in a specified volume of air. Referred to as a *control volume*, the model assumes that the predicted conditions within this volume are uniform at any time. Thus, the control volume has one temperature, smoke density, gas concentration, *etc*.

Different models divide the building into different numbers of control volumes depending on the desired level of detail. The most common fire model, known as a *zone model*, generally uses two control volumes to describe a compartment – an upper layer and a lower layer. In the compartment with the fire, additional control volumes for the fire plume or the ceiling jet may be included to improve the accuracy of the prediction (see Figure 1). Additional zones can be added as necessity arises to cover extensions.

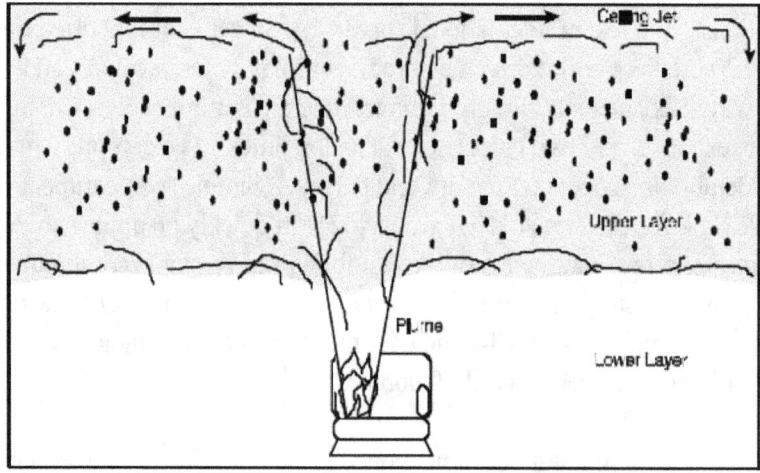

Figure 1. Zone model terms.

This two-layer approach has evolved from observation of such layering in real-scale fire experiments. Hot gases collect at the ceiling and fill the compartment from the top. While these experiments show some variation in conditions within the layer, these are small compared to the differences between the layers. Thus, the zone model can produce a fairly realistic simulation under most conditions.

Other types of models include *network models* and *field models*. Network models use one control volume per compartment and are used to predict conditions in spaces far removed from the fire compartment where temperatures are near ambient and layering does not occur. The field model goes to the other extreme, dividing the compartment into thousands or even a million or more control volumes. Such models can predict the variation in conditions within the layers, but typically require far longer run times than zone models. Thus, they are used when highly detailed calculations are essential.

1.1 Overview of the CFAST fire model

CFAST is a zone model used to calculate the evolving distribution of smoke, fire gases and temperature throughout a constructed facility during a fire. In CFAST, each compartment is divided into two layers.

The modeling equations used in CFAST take the mathematical form of an initial value problem for a system of ordinary differential equations (ODE). These equations are derived using the conservation of mass, the conservation of energy (equivalently the first law of thermodynamics), the ideal gas law and relations for density and internal energy. These equations predict as functions of time quantities such as pressure, layer heights and temperatures given the accumulation of mass and enthalpy in the two layers. The CFAST model then consists of a set of ODEs to compute the environment in each compartment and a collection of algorithms to compute the mass and enthalpy source terms required by the ODEs.

1.2 Implementation of the Conceptual Model

This section discusses each of the sub-models in CFAST. In general, the sections are similar to the way the model itself is structured. The sub-sections which follow discuss the way the actual phenomena are implemented numerically.

1.2.1 Fires

A fire in CFAST is implemented as a source of fuel which is released at a specified rate. This fuel is converted into enthalpy (the conversion factor is the heat of combustion) and mass (the conversion factor is the yield of a particular species) as it burns. A fire in CFAST is constrained if the enthalpy conversion depends on the oxygen concentration otherwise it is unconstrained. Burning can take place in the portion of the plume in the lower layer (if any), in the upper layer, or in a door jet. For an unconstrained fire, the burning will all take place within the fire plume. For a constrained fire, burning will take place where there is sufficient oxygen. When insufficient oxygen is entrained into the fire plume, unburned fuel will successively move into and burn in: the upper layer of the fire compartment, the plume in the doorway to the next compartment, the upper layer of the next compartment, the plume in the doorway to the third compartment, and so forth until it is consumed or gets to the outside.

This version of CFAST includes the ability to track, independently, multiple fires in one or more compartments of the building. These fires are treated as totally separate entities, i.e., with no interaction of the plumes or radiative exchange between fires in a compartment. These fires are generally referred to as "objects" and can be ignited at a specified time, temperature or heat flux.

This version of CFAST does not include a pyrolysis model to predict fire growth. Rather pyrolysis rates for each fire modeled define the fire history. The similarity of that input to the real fire problem of interest will determine the accuracy of the resulting calculation.

1.2.2 **Plumes and Layers**

A plume is formed above any burning object. It acts as a pump transferring mass and enthalpy from the lower layer into the upper layer. A correlation is used to predict the amount of mass and enthalpy that is transferred. A more complete plume model would predict plume entrainment by creating a separate zone and solving the appropriate equations.

Two sources exist for moving enthalpy and mass between the layers within and between compartments. Within the compartment, the fire plume provides one source. The other source of mixing between the layers occurs at vents such as doors or windows. Here, there is mixing at the boundary of the opposing flows moving into and out of the compartment. The degree of mixing is based on an empirically-derived mixing relation. Both the outflow and inflow entrain air from the surrounding layers. The flow at vents is also modeled as a plume (called the door plume or jet), and uses the same equations as the fire plume, with two differences. First, an offset is calculated to account for entrainment within the doorway and second, the equations are modified to account for the rectangular geometry of vents compared to the round geometry of fire plumes. All plumes within the simulation entrain air from their surroundings according to an empirically-derived entrainment relation. Entrainment of relatively cool, non-smoke laden air adds oxygen to the plume and allows burning of the fuel. It also causes it to expand as the plume moves upward in the shape of an inverted cone. The entrainment in a vent is caused by bi-directional flow and results from vortices formed near a shear layer. This phenomenon called the Kelvin-Helmholtz instability. It is not exactly the same as a normal plume, so some error arises when this entrainment is approximated by a normal plume entrainment algorithm.

While experiments show that there is very little mixing between the layers at their interface, sources of convection such as radiators or diffusers of heating and air conditioning systems, and the downward flows of gases caused by cooling at walls, will cause such mixing. These are examples of phenomena which are not included because the theories are still under development. Also, the plumes are *assumed* not to be affected by other flows which may occur. For example, if the burning object is near the door the strong inflow of air will cause the plume axis to lean away from the door and affect entrainment of gases into the plume. Such effects are not included in the model.

As discussed above, each compartment is divided into an upper and lower layer. At the start of the simulation, the layers in each compartment are initialized at ambient conditions and by default, the upper layer volume set to 0.001 of the compartment volume (an arbitrary, small value set to avoid the potential mathematical problems associated with dividing by zero). Other values can be set. As enthalpy and mass are pumped into the upper layer by the fire plume, the upper layer expands in volume causing the lower layer to decrease in volume and the interface to move downward. If the door to the next compartment has a soffit, there can be no flow through the vent from the upper layer until the interface reaches the bottom of that soffit. Thus in the early stages the expanding upper layer will push down on the lower layer air and force it into the next compartment through the vent by expansion.

Once the interface reaches the soffit level, a door plume forms and flow from the fire compartment to the next compartment is initiated. As smoke flow from the fire compartment fills the second compartment, the lower layer of air in the second compartment is pushed down. As a result, some of this air flows into the fire compartment through the lower part of the connecting doorway (or vent). Thus, a vent between the fire compartment and connecting compartments can have simultaneous, opposing flows of air. All flows are driven by pressure and density differences that result from temperature differences and layer depths. Thus the key to getting the right flows is to correctly distribute the fire and plume's mass and enthalpy between the layers.

1.2.3 Vent Flow

Flow through vents is a dominant component of any fire model because it is sensitive to small changes in pressure and transfers the greatest amount of enthalpy on an instantaneous basis of all the source terms (except of course for the fire and plume). Its sensitivity to environmental changes arises through its dependence on the pressure difference between compartments which can change rapidly.

CFAST models two types of vent flow, horizontal flow through vertical vents (ceiling holes, hatches *etc.*) and vertical flow through horizontal vents (doors, windows *etc.*). Horizontal flow is the flow which is normally thought of when discussing fires. Vertical flow is particularly important in two disparate situations: a ship, and the role of fire fighters doing roof venting.

Horizontal vent flow is determined using the pressure difference across a vent. Flow at a given elevation may be computed using Bernoulli's law by first computing the pressure difference at that elevation. The pressure on each side of the vent is computed using the pressure at the floor, the height of the floor and the density.

Atmospheric pressure is about 100 000 Pa, fires produce pressure changes from 1 Pa to 1000 Pa and mechanical ventilation systems typically involve pressure differentials of about 1 Pa to 100 Pa. The pressure variables are solved to a higher accuracy than other solution variables because of the subtraction (with resulting loss of precision) needed to calculate vent flows from pressure differences.

1.2.4 Heat Transfer

Gas layers exchange energy with their surroundings via convective and radiative heat transfer. Different material properties can be used for the ceiling, floor, and walls of each compartment (although all the walls of a compartment must be the same). Additionally, CFAST allows each surface to be composed of up to three distinct materials. This allows the user to deal naturally with the actual building construction. Material thermophysical properties are *assumed* to be constant, although we know that they actually vary with temperature. This assumption is made because data over the required temperature range is scarce even for common materials. However the user should recognize that the mechanical properties of some materials may change with temperature. These effects are not modeled.

Radiative transfer occurs among the fire(s), gas layers and compartment surfaces (ceiling, walls and floor). This transfer is a function of the temperature differences and the emissivity of the gas layers as well as the compartment surfaces. For the fire and typical surfaces, emissivity values only vary over a small range. For the gas layers, however, the emissivity is a function of the concentration of species which are strong radiators: predominately smoke particulates, carbon dioxide, and water. Thus errors in the species concentrations can give rise to errors in the distribution of enthalpy among the layers, which results in errors in temperatures, resulting in errors in the flows. This illustrates just how tightly coupled the predictions made by CFAST can be.

1.2.5 Species Concentration and Deposition

When the layers are initialized at the start of the simulation, they are set to ambient conditions. These are the initial temperatures specified by the user, and 23 % by mass (20.8 % by volume) oxygen, 77 % by mass (79 % by volume) nitrogen, a mass concentration of water specified by the user as a relative humidity, and a zero concentration of all other species. As fuel is pyrolyzed, the various species are produced in direct relation to the mass of fuel burned (this relation is the species yield specified by the user for the fuel burning). Since oxygen is consumed rather than produced by the burning, the "yield" of oxygen is negative, and is set internally to correspond to the amount of oxygen needed to burn the fuel. Also, hydrogen cyanide and hydrogen chloride are assumed to be products of pyrolysis whereas carbon dioxide, carbon monoxide, water, and soot are products of combustion.

Each unit mass of a species produced is carried in the flow to the various compartments and accumulates in the layers. The model keeps track of the mass of each species in each layer, and knows the volume of each layer as a function of time. The mass divided by the volume is the mass concentration, which along with the molecular weight gives the concentration in volume % or parts per million as appropriate.

CFAST uses a combustion chemistry scheme based on a carbon-hydrogen-oxygen balance. The scheme is applied in three places. The first is burning in the portion of the plume which is in the lower

6

layer of the compartment of fire origin. The second is the portion in the upper layer, also in the compartment of origin. The third is in the vent flow which entrains air from a lower layer into an upper layer in an adjacent compartment. This is equivalent to solving the conservation equations for each species independently.

2 Predictive Equations Used by the CFAST Model

This section presents a derivation of the predictive equations for zone fire models and explains in detail the ones used in CFAST [6], [8]. The zone fire model used in CFAST takes the form of an initial value problem for a mixed system of differential and algebraic equations. These equations are derived from the conservation of mass and energy. Subsidiary equations are the ideal gas law and definitions of density and internal energy (for example, see [11]). These conservation laws are invoked for each zone or control volume. For further information on the numerical implications of these choices please see reference [12].

The basic element of the model is a zone. The basic assumption of a zone model is that properties such as temperature can be approximated throughout the zone by some uniform function. The usual approximation is that temperature, density and so on are uniform within a zone. The assumption of uniform properties is reasonable and yields good agreement with experiment. In general, these zones are grouped within compartments.

There are two reasonable conjectures which dramatically improve the ease of solving these equations. Momentum is ignored within a compartment. The momentum of the interface has no significance in the present context. However, at boundaries such as windows, doors and so on, the Euler equation is integrated explicitly to yield the Bernoulli equation. This is solved implicitly in the equations which are discussed below. The other approximation is that the pressure is approximately uniform within a compartment. The argument is that a change in pressure of a few tens of Pascals over the height of the compartment is negligible in comparison with atmospheric pressure. Once again, this is applied to the basic conservation equations. This is consistent with the point source view of finite element models. Volume is merely one of the dependent variables. However, the hydrostatic variation in pressure *is* taken into account in calculating pressure differences between compartments.

Many formulations based upon these assumptions can be derived. Several of these are discussed later. One formulation can be converted into another using the definitions of density, internal energy and the ideal gas law. Though equivalent analytically, these formulations differ in their numerical properties. Also, until the development of FAST [6], all models of this type assumed that the pressure equilibrated instantaneously, and thus the *dP/dt* term could be set to zero. However, as has been shown [13], it is better to solve these equations in the differential rather than the algebraic form if the proper solver is used.

As discussed in references [12] and [14], the zone fire modeling differential equations (ODE's) are stiff. The term stiff means that multiple time scales are present in the ODE solution. In our problem, pressures adjust to changing conditions much quicker than other quantities such as layer temperatures or interface heights. Special solvers are required in general to solve zone fire modeling ODE's because of this stiffness. Runge-Kutta methods or predictor-corrector methods such as Adams-Bashforth require prohibitively small time steps in order to track the short-time scale phenomena (pressure in our case). Methods that calculate the Jacobian (or at least approximate it) have a much larger stability region for stiff problems and are thus more successful at their solution.

Each formulation can be expressed in terms of mass and enthalpy flow. These rates represent the exchange of mass and enthalpy between zones due to physical phenomena such as plumes, natural and forced ventilation, convective and radiative heat transfer, and so on. For example, a vent exchanges mass and enthalpy between zones in connected rooms, a fire plume typically adds heat to the upper layer and transfers entrained mass and enthalpy from the lower to the upper layer, and convection transfers enthalpy from the gas layers to the surrounding walls.

We use the formalism that the mass flow to the upper and lower layers is denoted $\dot{m}_U$ and $\dot{m}_L$ and the enthalpy flow to the upper and lower layers is denoted $\dot{s}_U$ and $\dot{s}_L$. It is tacitly assumed that these rates may be computed in terms of zone properties such as temperature and density. These rates represent the net sum of all possible sources of mass and enthalpy due to phenomena such as those listed above. The numerical characteristics of the various formulations are easier to identify if the underlying physical phenomena are decoupled in this way.

Many approximations are necessary when developing physical sub-models for the mass and enthalpy terms. For example, most fire models assume that 1) the specific heat terms c_p and c_v are constant even though they depend upon temperature, 2) hydrostatic terms can be ignored in the equation of state (the ideal gas law) relating density of a layer with its temperature. However, the derivations which follow are all based on the basic conservation laws.

2.1 Derivation of Equations for a Two-Layer Model

A compartment is divided into two control volumes, a relatively hot upper layer and a relatively cooler lower layer, as illustrated in Figure 2. The gas in each layer has attributes of mass, internal energy, density, temperature, and volume denoted respectively by m_i, E_i, ρ_i, T_i, and V_i where $i=L$ for the lower layer and $i=U$ for the upper layer. The compartment as a whole has the attribute of pressure P. These 11 variables are related by means of the following seven constraints

$$\rho_i = \frac{m_i}{V_i} \qquad \text{(density)} \tag{1}$$

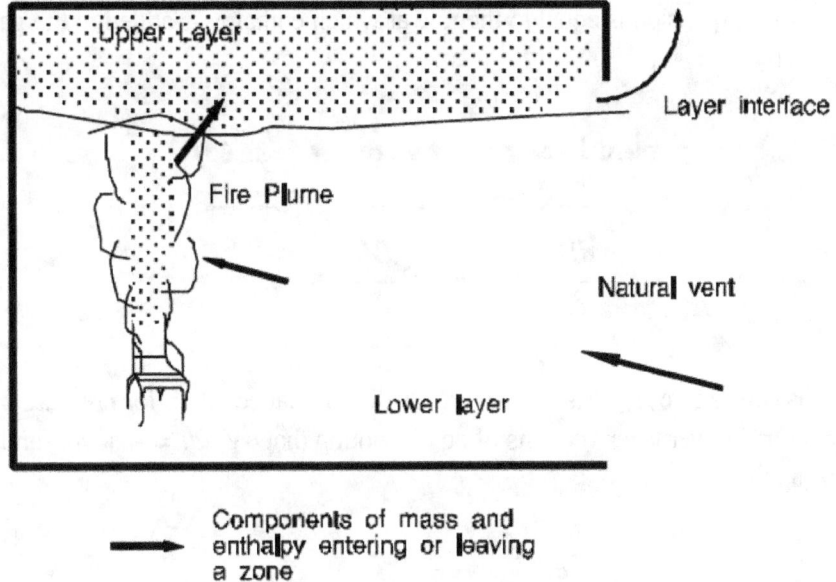

Figure 2. Schematic of control volumes in a two-layer zone model.

$$E_i = c_v m_i T_i \quad \text{(internal energy)} \tag{2}$$

$$P = R\rho_i T_i \quad \text{(ideal gas law)} \tag{3}$$

$$V = V_L + V_U \quad \text{(total volume)} \tag{4}$$

We get seven by counting density, internal energy and the ideal gas law twice (once for each layer). The specific heat at constant volume and at constant pressure c_v and c_p, the universal gas constant, R, and the ratio of specific heats, γ, are related by $\gamma = c_p / c_v$ and $R = c_p - c_v$. For air, $c_p \approx 1000$ kJ/kg K and $\gamma = 1.4$. Four additional equations obtained from conservation of mass and energy for each layer are required to complete the equation set. The differential equations for mass in each layer are

$$\frac{dm_L}{dt} = \dot{m}_L$$
$$\frac{dm_U}{dt} = \dot{m}_U \tag{5}$$

11

The first law of thermodynamics states that the rate of increase of internal energy plus the rate at which the layer does work by expansion is equal to the rate at which enthalpy is added to the gas. In differential form this is

$$
\underbrace{\frac{dE_i}{dt}}_{\text{internal energy}} + \underbrace{P\frac{dV_i}{dt}}_{\text{work}} = \underbrace{\dot{\varepsilon}_i}_{\text{enthalpy}} \tag{6}
$$

where c_v is taken as constant, $c_p/c_v = \gamma$ and $c_p - c_v = R$. A differential equation for pressure can be derived by adding the upper and lower layer versions of eq (6), noting that $dV_U/dt = -dV_L/dt$, and substituting the differential form of eq (2) to yield

$$
\frac{dP}{dt} = \frac{\gamma - 1}{V}\left(\dot{\varepsilon}_L + \dot{\varepsilon}_U\right) \tag{7}
$$

Differential equations for the layer volumes can be obtained by substituting the differential form of eq (2) into eq (6) to obtain

$$
\frac{dV_i}{dt} = \frac{1}{P\gamma}\left((\gamma - 1)\,\dot{\varepsilon}_i - V_i\frac{dP}{dt}\right). \tag{8}
$$

Equation (6) can be rewritten using eq (8) to eliminate dV/dt to obtain

$$
\frac{dE_i}{dt} = \frac{1}{\gamma}\left(\dot{\varepsilon}_i + V_i\frac{dP}{dt}\right). \tag{9}
$$

A differential equation for density can be derived by applying the quotient rule to $\dfrac{d\rho_i}{dt} = \dfrac{d}{dt}\left(\dfrac{m_i}{V_i}\right)$ and using eq (8) to eliminate dV_i/dt to obtain

$$
\frac{d\rho_i}{dt} = -\frac{1}{c_p T_i V_i}\left((\dot{\varepsilon}_i - c_p \dot{m}_i T_i) - \frac{V_i}{\gamma - 1}\frac{dP}{dt}\right). \tag{10}
$$

12

Temperature differential equations can be obtained from the equation of state by applying the quotient rule to $\frac{dT_i}{dt} = \frac{d}{dt}\left(\frac{P}{R\rho_i}\right)$ and using eq (10) to eliminate $d\rho/dt$ to obtain

$$\frac{dT_i}{dt} = \frac{1}{c_p \rho_i V_i}\left(\left(\dot{s}_i - c_p \dot{m}_i T_i\right) + V_i \frac{dP}{dt}\right).$$ (11)

These equations for each of the eleven variables are summarized in table 1. The time evolution of these solution variables can be computed by solving the corresponding differential equations together with appropriate initial conditions. The remaining seven variables can be determined from the four solution variables using eqs (1) to (4).

There are, however, many possible differential equation formulations. Indeed, there are 330 different ways to select four variables from eleven. Many of these systems are incomplete due to the relationships that exist between the variables given in eqs (1) to (4). For example the variables, ρ_U, V_U, m_U, and P form a dependent set since $\rho_U = m_U / V_U$. Table 2 shows the solution variable selection made by several zone fire models.

The number of differential equation formulations can be considerably reduced by not mixing variable types between layers; that is, if upper layer mass is chosen as a solution variable, then lower layer mass must also be chosen. For example, for two of the solution variables choose m_L and m_U, or ρ_L and ρ_U, or T_L and T_U. For the other two solution variables pick E_L and E_U or P and V_L or P and V_U. This reduces the number of distinct formulations to nine. Since the numerical properties of the upper layer volume equation are the same as a lower layer one, the number of distinct formulations can be reduced to six.

13

Table 1. Conservative zone model equations

Equation Type	Differential Equation
i'th layer mass	$$\frac{dm_i}{dt} = \dot{m}_i$$
pressure	$$\frac{dP}{dt} = \frac{\gamma-1}{V}\left(\dot{s}_L + \dot{s}_U\right)$$
i'th layer energy	$$\frac{dE_i}{dt} = \frac{1}{\gamma}\left(\dot{s}_i + V_i\frac{dP}{dt}\right)$$
i'th layer volume	$$\frac{dV_i}{dt} = \frac{1}{\gamma P}\left((\gamma-1)\dot{s}_i - V_i\frac{dP}{dt}\right)$$
i'th layer density	$$\frac{d\rho_i}{dt} = -\frac{1}{c_p T_i V_i}\left((\dot{s}_i - c_p\dot{m}_i T_i) - \frac{V_i}{\gamma-1}\frac{dP}{dt}\right)$$
i'th layer temperature	$$\frac{dT_i}{dt} = \frac{1}{c_p \rho_i V_i}\left((\dot{s}_i - c_p\dot{m}_i T_i) + V_i\frac{dP}{dt}\right)$$

Table 2. Conservative zone model equation selections

Zone Fire Model	Equations	Substitutions
FAST	$\dfrac{d\Delta P}{dt}$, $\dfrac{dV_L}{dt}$, $\dfrac{dT_U}{dt}$, $\dfrac{dT_L}{dt}$	$\Delta P = P - P_{ref}$
CCFM.HOLE	$\dfrac{d\Delta P}{dt}$, $\dfrac{dy}{dt}$, $\dfrac{d\rho_U}{dt}$, $\dfrac{d\rho_L}{dt}$	$\Delta P = P - P_{ref}$ $y = V_L / A_{room}$
CCFM.VENTS	$\dfrac{d\Delta P}{dt}$, $\dfrac{dy}{dt}$, $\dfrac{dm_U}{dt}$, $\dfrac{dm_L}{dt}$	$\Delta P = P - P_{ref}$ $y = V_L / A_{room}$
FIRST, HARVARD	$\dfrac{dE_U}{dt}$, $\dfrac{dE_L}{dt}$, $\dfrac{dm_U}{dt}$, $\dfrac{dm_L}{dt}$	

2.2 Equation Set Used in CFAST

The current version of CFAST is set up to use the equation set for layer temperature, layer volume, and pressure as shown below. However, the internal structure of the model is such that it will allow any of the formulations above to be substituted with relative ease.

$$P = P_{ref} + \Delta P \tag{12}$$

$$\frac{dP}{dt} = \frac{\gamma - 1}{V}\left(\dot{s}_L + \dot{s}_U\right) \tag{13}$$

$$\frac{dV_U}{dt} = \frac{1}{\gamma P}\left((\gamma - 1)\dot{s}_U - V_U\frac{dP}{dt}\right) \tag{14}$$

$$\frac{dT_U}{dt} = \frac{1}{c_p\rho_U V_U}\left((\dot{s}_U - c_p\dot{m}_U T_U) + V_U\frac{dP}{dt}\right) \tag{15}$$

$$\frac{dT_L}{dt} = \frac{1}{c_p\rho_L V_L}\left((\dot{s}_L - c_p\dot{m}_L T_L) + V_L\frac{dP}{dt}\right) \tag{16}$$

15

3 Source Terms for the CFAST Model

The conserved quantities in each compartment are described by the set of predictive equations above. The form of the equations is such that the physical phenomena are source terms* on the right-hand-side of these equations. Such a formulation makes the addition and deletion of physical phenomena and changing the form of algorithms a *relatively* simple matter. For each of the phenomena discussed below, the physical basis for the model is discussed first, followed by a brief presentation of the implementation within CFAST. For all of the phenomena, there are basically two parts to the implementation: the physical interface routine (which is the interface between the CFAST model and the algorithm) and the actual physical routine(s) which implement the physics. This implementation allows the physics to remain independent of the structure of CFAST and allows easier insertion of new phenomena.

3.1 The Fire

3.1.1 Specified Fire (Fire Types 1 and 2)

A specified fire is one for which the time dependent characteristics are specified as a function of time. The specified fire can be unconstrained (type 1) or constrained (type 2). The heat release rate for a constrained fire may be reduced below its specified value based upon the concentration of fuel or oxygen available for combustion. Combustion chemistry is not calculated for type 1 fires. The pyrolysis rate for both fire types is specified as $\dot{m}_f$, the burning rate as $\dot{m}_b$ and the heat of combustion as H_c so that the heat release rate, Q_f, is

$$Q_f = H_c \dot{m}_b - c_p \left(T_u - T_v \right) \dot{m}_b .$$ (17)

For the unconstrained fire, $\dot{m}_b = \dot{m}_f$, whereas for the constrained fire, $\dot{m}_b < \dot{m}_f$, or equivalently the burning rate may be less than the pyrolysis rate. Models of specified fires generally use an effective heat of combustion which is obtained from an experimental apparatus such as the cone calorimeter [15]. A shortcoming of this approach is that it does not account for increased pyrolysis due to radiative feedback from the flame or compartment. In an actual fire, this is an important consideration, and the specification used should match the experimental conditions as closely as possible.

The enthalpy which is released goes into radiation and convection

---

*The $\dot{m}$ and $\dot{s}$ in eqs (13) to (16)

17

$$\dot{Q}_r (fire) = \chi_R \dot{Q}_f$$
$$\dot{Q}_c (fire) = (1-\chi_R)\dot{Q}_f .$$

(18)

where, χ_R, is the fraction of the fire's heat release rate given off as radiation. The convective heat release rate, $\dot{Q}_c (fire)$ then becomes the driving term in the plume flow. For a specified fire there is radiation to both the upper and lower layers, whereas the convective part contributes only to the upper layer.

3.1.2 Combustion Chemistry (Fire Type 2)

Two types of fires can be selected. The first generates heat as described above. The second type of fire is constrained by the amount of available oxygen. The latter scheme is applied in three places. The first is burning in the portion of the plume which is in the lower layer of the room of fire origin (region #1). The second is the portion of the plume in the upper layer, also in the room of origin (region #2). The third is in the vent flow which entrains air from a lower layer into an upper layer in an adjacent compartment (region #3).These are shown schematically in Figure 3.

The species which are affected by this scheme are O_2, CO_2, CO, H_2O, unburned hydrocarbons (TUHC), and soot (OD). Nitrogen is carried as a gas, but only acts as a diluent. There are at present no nitrogen reactions. In a chemical equation, the individual atoms on the left and right hand sides must balance. This is true regardless of whether the reaction is considered to be stoichiometric (complete). We apply this idea to the combination of fuel and oxygen to yield a balance of number density (#/volume). In terms of the "regions," (Figure 3), we have

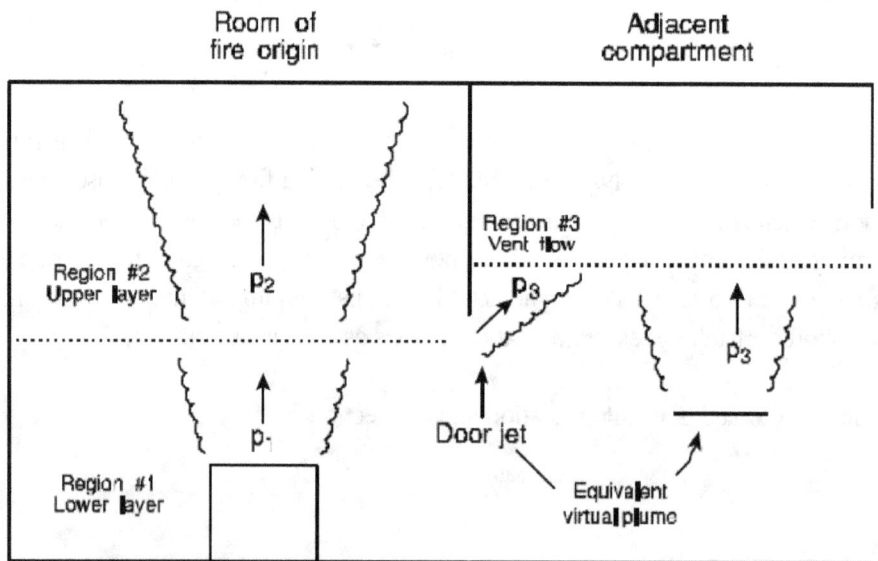

Figure 3. Schematic of entrainment and burning regions.

18

$$\dot{m}_f = \text{pyrolysis rate of the source (kg/sec) (region \#1)}$$

or

$$\dot{m}_f = \dot{m}_{tuhc} \text{ from a previous region (kg/sec) (region \#2 and \#3).}$$

and

$$\dot{m}_{tuhc} = \dot{m}_f - \dot{m}_b$$

where *tuhc* stands for total un-burned hydrocarbons.

The simplest form of energy release is made by specifying a heat release rate, together with a consistent mass release rate. This would simulate the fire that occurs in an unconfined space. As soon as one is constrained by the confines of a compartment, then the nature of the fire changes. In particular the available oxygen may not be sufficient to allow complete combustion. However, it is not consistent to try to account for the oxygen alone. All pertinent species must be followed.

The essence of the species production scheme which we now utilize is to allow as realistic fuel composition as possible, i.e., include oxygen, carbon, hydrogen and chlorine as part of the fuel. Carbon monoxide, carbon dioxide, soot, water, hydrogen cyanide and hydrogen chloride are the products of combustion. The fuel properties are specified as H/C, HCl/C, HCN/C and O/C which are mass ratios of hydrogen, hydrochloric acid, hydrogen cyanide and oxygen to carbon respectively. The production properties are HCl/f, HCN/f, CO/CO_2, and S/CO_2 which again are mass ratios. The chemical symbols used here have their usual meaning, except for soot. The subscript "S" is used to designate soot, and we assume it consists primarily of carbon, at least by mass.

The fuel burning rate in terms of the carbon production is

$$\dot{m}_f = \{-\} \times \dot{m}_c \tag{19}$$

where $\{-\}$ is the multiplier in the fuel production defined as

$$\{-\} = \left(1 + \frac{H}{C} + \frac{HCl}{C} + \frac{HCN}{C} + \frac{O}{C} \right) = f/C. \tag{20}$$

In order to avoid detailed chemical kinetics, we use the oxygen consumption concept [16], [17] to relate the mass loss to the heat release rate. The following derivation is for the heat release rate as a

19

function of the fuel burning rate, and the heat release rate based on oxygen consumption. H/C, HCl/C, HCN/C and O/C are the ratio of mass of that species to carbon in the fuel. Thus H/C is for the mass of hydrogen to the mass of carbon produced in pyrolysis. This is a very useful way to characterize the fuel. This is in terms of the elemental composition of the fuel, and not elemental molecules, such as H_2. These are the ratios for the fuel, and the material which comes from it. For the products of the combustion process, we have CO_2/C, CO/C, H_2O/C and S/C. These ratios are in terms of free molecules, generally gaseous.

The first step is to limit the actual burning which takes place in the combustion zone. In each combustion zone, there is a quantity of fuel available. At the source this results from the pyrolysis of the material, m_f. In other situations such as a plume or door jet, it is the net unburned fuel available, m_{TUHC}. In each case, the fuel which is available but not burned is then deposited into the "m_{TUHC}" category. This provides a consistent notation. In the discussion below, m_f is the amount of fuel burned. This value is initialized to the available fuel, and then reduced if there is insufficient oxygen to support complete combustion. Subsequently, the available fuel, m_{TUHC}, is reduced by the final value of m_f. Thus we have a consistent description in each burning region, with an algorithm that can be invoked independent of the region being analyzed.

$$\dot{Q} = \dot{m}_f \times H_c \, . \tag{21}$$

with the mass of oxygen required to achieve this energy release rate (based on the oxygen consumption principle [18]) of

$$\dot{m}_O = \frac{\dot{Q}}{1.32 \times 10^7} = \dot{m}_f \times \frac{H_c}{1.32 \times 10^7} \, . \tag{22}$$

If the fuel contains oxygen (available for combustion), the oxygen needed to achieve full combustion is less than this value

$$\dot{m}_O(\text{needed}) = \dot{m}_O - \dot{m}_O(\text{in the fuel}) \tag{23}$$

If sufficient oxygen is available, then it is fully burned. However, if the oxygen concentration is low enough, it will constrain the burning and impose a limit on the amount of fuel actually burned, as opposed to the amount pyrolyzed. The actual limitation is discussed below and is presented as eq (38).

$$\dot{m}_O(\text{actual}) = \text{minimum of} \left\{ \dot{m}_O(\text{available}), \dot{m}_O(\text{needed}) \right\}. \tag{24}$$

$$\dot{m}_f(\text{actual}) = \dot{m}_O(\text{actual}) \times \frac{1.32 \times 10^7}{H_c} \tag{25}$$

Essentially, we limit the amount of fuel that is burned, as opposed to the amount that is pyrolyzed, to the lesser of the amount pyrolyzed and that required to consume the *available* oxygen. The $\dot{m}_O$(actual) and $\dot{m}_f$(actual) are the quantities used below. By way of explanation, eq (21) tells us how much energy would be released by the available fuel if there were no constraint (free burn). Equation (22) then tells us the mass of oxygen required to achieve this energy release rate. The relationship is based on the work in reference [19]. Equation (23) yields the amount needed based on the required amount less the oxygen available in the fuel. Solid propellant would yield a value of zero at this point. Equation (24) limits the amount used and eq (25) then yields the amount of fuel actually burned, as opposed to the amount pyrolyzed.

We begin with the mass balance equation. The mass consumed as pyrolyzate plus oxygen must reappear as product.

$$\dot{m}_f + \dot{m}_O = \dot{m}_f + \dot{m}_f \times \frac{H_c}{1.32 \times 10^7} - \frac{\dot{m}_f}{\{-\}} \times \left(\frac{O}{C} \right)$$

$$= \dot{m}_{CO_2} + \dot{m}_{CO} + \dot{m}_S + \dot{m}_{H_2O} + \dot{m}_{HCl} + \dot{m}_{HCN}$$

(26)

We then substitute the following definitions of mass produced of each species based on the amount of carbon (ala. fuel) consumed as

$$\dot{m}_{HCl} = \left(\frac{HCl}{C} \right) \times \dot{m}_C \rightarrow \left(\frac{HCl}{f} \right) \times \dot{m}_f$$

(27)

$$\dot{m}_{HCN} = \left(\frac{HCN}{C} \right) \times \dot{m}_C \rightarrow \left(\frac{HCN}{f} \right) \times \dot{m}_f$$

(28)

$$\dot{m}_{H_2O} = \frac{1}{2} \left(\frac{H_2O}{H} \right) \times \left(\frac{H}{C} \right) \times \dot{m}_C = 9 \times \left(\frac{H}{C} \right) \times \dot{m}_C \rightarrow 9 \times \left(\frac{H}{C} \right) \times \frac{\dot{m}_f}{\{-\}}$$

(29)

$$\dot{m}_{CO_2} = \left(\frac{CO_2}{C} \right) \times \dot{m}_C$$

(30)

$$\dot{m}_S = \left(\frac{S}{C} \right) \times \dot{m}_C = \left(\frac{CO_2}{C} \right) \times \left(\frac{S}{CO_2} \right) \times \dot{m}_C \rightarrow \left(\frac{S}{CO_2} \right) \times \dot{m}_{CO_2}$$

(31)

$$\dot{m}_{CO} = \left(\frac{CO}{C}\right) \times \dot{m}_C = \left(\frac{CO_2}{C}\right) \times \left(\frac{CO}{CO_2}\right) \times \dot{m}_C \rightarrow \left(\frac{CO}{CO_2}\right) \times \dot{m}_{CO_2} \tag{32}$$

Substituting the above definitions into the mass balance equation yields:

$$\left(\frac{CO_2}{C}\right) = \frac{\{-\} \times \left(1 + \dfrac{H_e}{1.32 \times 10^7} - \dfrac{O/C}{\{-\}}\right) - \left(\dfrac{HCl}{C} + \dfrac{HCN}{C} + 9\dfrac{H}{C}\right)}{\left(1 + \dfrac{S}{CO_2} + \dfrac{CO}{CO_2}\right)} \tag{33}$$

With these definitions, we can substitute back into the equation for carbon dioxide production, which yields

$$\dot{m}_{CO_2} = \dot{m}_f \times \frac{\left(1 + \dfrac{h_e}{1.32 \times 10^7} - \dfrac{O/C}{\{-\}}\right) - \dfrac{\left(\dfrac{HCl}{C} + \dfrac{HCN}{C} + 9\dfrac{H}{C}\right)}{\{-\}}}{\left(1 + \dfrac{S}{CO_2} + \dfrac{CO}{CO_2}\right)}. \tag{34}$$

and the remainder follow explicitly.

The form in which we cast these equations evolves naturally from the properties of combustion. Hydrogen, carbon and bound oxygen are properties of the fuel. They can be measured experimentally independent of the combustion process. Thus we use these ratios as the basis of the scheme. In a similar sense, hydrogen chloride and hydrogen cyanide are properties of the pyrolysis process. So hydrogen chloride and hydrogen cyanide production are specified with respect to the fuel pyrolysis. Normally this is how they are measured, for example with the cone calorimeter, so we can use the measured quantities directly. Other than the cyanide, chloride and water production, hydrogen does not play a role. In general, hydrogen has a much greater affinity for oxygen than carbon, so almost all of the hydrogen will be utilized. This dictates our next choice, which is that soot is essentially all carbon. On a mass basis, this is certainly true. On a molecular basis, however, it may not be so simple. Carbon dioxide is a direct product of combustion, and the assumption is that most carbon will end up here. Carbon monoxide and soot are functions of incomplete combustion. Thus they depend on the environment in which the burning takes place. They are in no case a function of the pyrolysis process itself. Thus the production of these products is specified with respect to the carbon dioxide. At present, we must rely on measured ratios, but this is beginning to change as we gain a better understanding of the combustion process. So, in the present model, carbon goes to one of three final species, carbon dioxide, carbon monoxide or soot, with the particular branching ratio depending on the chemistry active at the time.

Eqs (29) through (34) are used in terms of the carbon production. We now need to recast HCl and HCN in terms of fuel production rather than carbon production, since that is how they are measured. Since HCl and HCN are similar, we will just make the argument for one, and then assume that the derivation is the same for the other. One simplification will be possible for the HCN though, and that is that its production rate is *always* much less than the pyrolysis rate.

Since {—} is just f/C,

$$\left(\frac{HCl}{C}\right) = \left(\frac{HCl}{f}\right) \times \left(1 + \frac{H}{C} + \frac{HCl}{C} + \frac{HCN}{C} + \frac{O}{C}\right). \tag{35}$$

Therefore

$$\left(\frac{HCl}{C}\right) = \left(\frac{HCl}{f}\right) \times \left(\frac{1 + \frac{H}{C} + \frac{O}{C}}{1 - \left(\frac{HCl}{f}\right)}\right), \tag{36}$$

and for hydrogen cyanide we have

$$\left(\frac{HCN}{C}\right) = \left(\frac{HCN}{f}\right) \times \left(1 + \frac{H}{C} + \frac{HCl}{C} + \frac{O}{C}\right). \tag{37}$$

In this latter case, we assume that the cyanide ratio (HCN/C) is small compared to unity. It is the HCl/C and HCN/C ratios which are used by the model.

The relationship between oxygen and fuel concentration defines a range where burning will take place. The rich limit is where, for a given ratio of O_2 to N_2 (generally the ratio in air), there is too much fuel for combustion. At the other end, there is the lean flammability limit, where there is too little fuel for combustion. In the CFAST model, the rich limit is incorporated by limiting the burning rate as the oxygen level decreases until a "lower oxygen limit" (LOL) is reached. The lower oxygen limit is incorporated through a smooth decrease in the burning rate near the limit:

$$\dot{m}_o(available) = \dot{m}_s Y_{O_2} C_{LOL} \tag{38}$$

where m_e is the mass entrainment flow rate and the lower oxygen limit coefficient, C_{LOL}, is the fraction of the available fuel which can be burned with the available oxygen and varies from 0 at the limit to 1 above the limit. The functional form provides a smooth cutoff of the burning over a narrow range above the limit.

$$ C_{LOL} = \frac{\tanh\left(800\left(Y_{O_2} - Y_{LOL}\right) - 4\right) + 1}{2} \tag{39} $$

For the lean flammability limit, an ignition temperature criterion is included, below which no burning takes place.

As stated, the burning rate simply decreases as the oxygen level decreases. We know that there is an oxygen concentration below which fuel will not oxidize. This is referred to as the "rich flammability" limit. In the present context we refer to this point as the lower oxygen limit (LOL). At the other end, there is a "lean flammability" limit. The fuel oxidation rate is limited at both ends. At present, we have incorporated only the rich flammability limit. We do not have sufficient theoretical underpinnings, nor sufficient experimental data, to include temperature dependence or the lean flammability limit. In the lean flammability limit, we use only a temperature criterion below which we assume no burning takes place.

In summary, we can predict the formation of some of the products of combustion, carbon dioxide, carbon monoxide, soot, water, hydrogen cyanide, and hydrogen chloride given the branching ratios CO/CO_2, $S(soot)/CO_2$, the composition of the fuel, H/C, O/C, HCl/f and HCN/f and the flammability limit. At present, in practice we use experimental values, such as those from Morehart et al.[20]. The composition of the fuel is a measurable quantity, although it is complicated somewhat by physical effects. The complication arises in that materials such as wood will yield methane in the early stages of burning, and carbon rich products at later times. Thus the H/C and O/C ratios are functions of time. Finally, the production ratios of CO/CO_2, $S(soot)/CO_2$ are based on the kinetics which in turn is a function of the ambient environment.

In earlier versions of CFAST, the chemistry routine was called once. However, a recent test case pointed out a long standing problem in the way CFAST coordinates plume entrainment with fire size. Before the fix outlined here, CFAST calculated plume entrainment via a two step process:

1) Determine the plume entrainment in FIRPLM via McCaffrey's method using the fire size unconstrained by the available oxygen.

2) Once the actual fire size is calculated from the available oxygen in CHEMIE, a new estimate for the plume entrainment is determined by a simple linear correction of

$$\dot{m}_{o,actual} = \dot{m}_{o,unconstrained} \frac{\dot{Q}_{actual}}{\dot{Q}_{unconstrained}} \qquad (40)$$

Since the plume entrainment is not a linear function of the fire size and the fire size depends on the oxygen entrained, this simple process can lead to an inconsistency between the fire size (calculated from the unconstrained plume entrainment) and the new estimate of the plume entrainment. For very large fires where the fire size is limited by the amount of oxygen entrained, this can lead to significant differences between the calculated fire size and the amount of oxygen actually available for the combustion.

The fix is a simple one: when the fire is limited by the available oxygen entrained into the plume, the plume entrainment and fire size are both re-calculated by calling FIRPLM and CHEMIE a second time to get a better estimate of the actual oxygen available (and thus the actual fire size).

3.2 Plumes

Buoyancy generated by the combustion processes in a fire causes the formation of a plume. Such a plume can transport mass and enthalpy from the fire into the lower or upper layer of a compartment. In the present implementation, we assume that both mass and enthalpy from the fire are deposited only into the upper layer. In addition the plume entrains mass from the lower layer and transports it into the upper layer. This yields a net enthalpy transfer between the two layers.

A fire generates energy at a rate Q. Some fraction, χ_R, will exit the fire as radiation. The remainder, χ_C, will then be deposited in the layers as convective energy or heat additional fuel so that it pyrolyzes. We can use the work of McCaffrey [21] to estimate the mass entrained by the fire/plume from the lower into the upper layer. This correlation divides the flame/plume into three regions as given in eq(41). This prescription agrees with the work of Cetegen et al. [22] in the intermittent regions but yields greater entrainment in the other two regions. This difference is particularly important for the initial fire since the upper layer is far removed from the fire.

$$
\begin{array}{llll}
\text{flaming:} & \dfrac{\dot{m}_e}{\dot{Q}} = 0.011 \left(\dfrac{z}{\dot{Q}^{2/5}} \right)^{0.566} & 0.00 \le \left(\dfrac{z}{\dot{Q}^{2/5}} \right) < 0.08 & \\[3ex]
\text{intermittent:} & \dfrac{\dot{m}_e}{\dot{Q}} = 0.026 \left(\dfrac{z}{\dot{Q}^{2/5}} \right)^{0.909} & 0.08 \le \left(\dfrac{z}{\dot{Q}^{2/5}} \right) < 0.20 & (41) \\[3ex]
\text{plume:} & \dfrac{\dot{m}_e}{\dot{Q}} = 0.124 \left(\dfrac{z}{\dot{Q}^{2/5}} \right)^{1.895} & 0.20 \le \left(\dfrac{z}{\dot{Q}^{2/5}} \right) &
\end{array}
$$

25

McCaffrey's correlation is an extension of the common point source plume model, with a different set of coefficients for each region. These coefficients are experimental correlations, and are not based on theory.

Within CFAST, the radiative fraction defaults to 0.30 [23]; i.e., 30 % of the fires energy is released via radiation. For other fuels, the work or Tewarson [24], McCaffrey [25], or Koseki [26] is available for reference. These place the typical range for the radiative fraction from about 0.15 to 0.5.

In CFAST, there is a constraint on the quantity of gas which can be entrained by a plume arising from a fire. The constraint arises from the physical fact that a plume can rise only so high for a given size of a heat source. In the earlier versions of this model (FAST version 17 and earlier), the plume was not treated as a separate zone. Rather we assumed that the upper layer was connected immediately to the fire by the plume. The implication is that the plume is formed instantaneously and stretches from the fire to the upper layer or ceiling. Consequently, early in a fire, when the energy flux was very small and the plume length very long, the entrainment was over predicted. This resulted in the interface falling more rapidly than was seen in experiments. Also the initial temperature was too low and the rate of rise too fast, whereas the asymptotic temperature was correct. The latter occurred when these early effects were no longer important.

The correct sequence of events is for a small fire to generate a plume which does not reach the ceiling or upper layer initially. The plume entrains enough cool gas to decrease the buoyancy to the point where it no longer rises. When there is sufficient energy present in the plume, it will penetrate the upper layer. The effect is two-fold: first, the interface will take longer to fall and second, the rate of rise of the upper layer temperature will not be as great. To this end the following prescription has been incorporated: for a given size fire, a limit is placed on the amount of mass which can be entrained, such that no more is entrained than would allow the plume to reach the layer interface. The result is that the interface falls at about the correct rate, although it starts a little too soon, and the upper layer temperature is over predicted, but follows experimental data after the initial phase (see sec. 5).

For the plume to be able to penetrate the inversion formed by a hot gas layer over a cooler gas layer, the density of the gas in the plume at the point of intersection must be less than the density of the gas in the upper layer. In practice, this places a maximum on the air entrained into the plume. From conservation of mass and enthalpy, we have

$$\dot{m}_p = \dot{m}_f + \dot{m}_e \tag{42}$$

$$\dot{m}_p c_p T_p = \dot{m}_f c_p T_f + \dot{m}_e c_p T_l \tag{43}$$

where the subscripts p, f, e, and l refer to the plume, fire, entrained air, and lower layer, respectively.

The criterion that the density in the plume region be lower than the upper layer implies that $T_u < T_p$. Solving eq (43) for T_p and eliminating $\dot{m}_p$ using eq (42) yields

$$T_p = \frac{T_f \dot{m}_f + T_l \dot{m}_e}{\dot{m}_f + \dot{m}_e} > T_u \tag{44}$$

or

$$\dot{m}_e < \left(\frac{T_f - T_u}{T_u - T_l}\right)\dot{m}_f < \frac{T_f}{T_u - T_l}\dot{m}_f \tag{45}$$

Substituting the convective energy released by the fire,

$$\dot{Q}_c(fire) = \dot{m}_f c_p T_f , \tag{46}$$

Substituting eq (46) into eq (45) yields the final form of the entrainment limit used in the CFAST model:

$$\dot{m}_e < \frac{\dot{Q}_c(fire)}{c_p(T_u - T_l)} \tag{47}$$

which is incorporated into the model. It should be noted that both the plume and layers are assumed to be well mixed with negligible mixing and transport time for the plume and layers.

3.3 Vent Flow

Mass flow (in the remainder of this section, the term "flow" will be used to mean mass flow) is the dominant source term for the predictive equations because it fluctuates most rapidly and transfers the greatest amount of enthalpy on an instantaneous basis of all the source terms (except of course the fire). Also, it is most sensitive to changes in the environment. CFAST models horizontal flow through vertical vents and vertical flow through horizontal vents. Horizontal flow encompasses flow through doors, windows and so on. Horizontal flow is discussed in section 3.3.1. Vertical flow occurs in ceiling vents. It is important in two separate situations: on a ship with open hatches and in house fires with roof venting. Vertical flow is discussed in section 3.3.2.

3.3.1 Horizontal Flow Through Vertical Vents

Flow through normal vents such as windows and doors is governed by the pressure difference across a vent. A momentum equation for the zone boundaries is not solved directly. Instead momentum transfer at the zone boundaries is included by using an integrated form of Euler's equation, namely Bernoulli's solution for the velocity equation. This solution is augmented for restricted openings by using flow coefficients [27] to allow for constriction from finite size doors. The flow (or orifice) coefficient is an empirical term which addresses the problem of constriction of velocity streamlines at an orifice.

Bernoulli's equation is the integral of the Euler equation and applies to general initial and final velocities and pressures. The implication of using this equation for a zone model is that the initial velocity in the doorway is the quantity sought, and the final velocity in the target compartment vanishes. That is, the flow velocity vanishes where the final pressure is measured. Thus, the pressure at a stagnation point is used. This is consistent with the concept of uniform zones which are completely mixed and have no internal flow. The general form for the velocity of the mass flow is given by

$$v = C\left(\frac{2\Delta P}{\rho}\right)^{1/2} \tag{48}$$

where C is the constriction (or flow) coefficient (≈ 0.7), ρ is the gas density on the source side, and ΔP is the pressure across the interface. (Note: at present we use a constant C for all gas temperatures)

The simplest means to define the limits of integration is with neutral planes, that is the height at which flow reversal occurs, and physical boundaries such as sills and soffits. By breaking the integral into intervals defined by flow reversal, a soffit, a sill, or a zone interface, the flow equation can be integrated piecewise analytically and then summed.

The approach to calculating the flow field is of some interest. The flow calculations are performed as follows. The vent opening is partitioned into at most six slabs where each slab is bounded by a layer height, neutral plane, or vent boundary such as a soffit or sill. The most general case is illustrated in Figure 4.

The mass flow for each slab can be determined from

$$\dot{m}_{t \to o} = \frac{1}{3}C(8\rho)A_{slab}\left(\frac{x^2 + xy + y^2}{x + y}\right) \tag{49}$$

where $x = |P_t|^{1/2}$, and $y = |P_b|^{1/2}$. P_t and P_b are the cross-vent pressure differential at the top and bottom of the slab respectively and A_{slab} is the cross-sectional area of the slab. The value of the density, ρ, is taken from the source compartment.

A mixing phenomenon occurs at vents which is similar to entrainment in plumes. As hot gases from one compartment leave that compartment and flow into an adjacent compartment a door jet can exist which is analogous to a normal plume. Mixing of this type occurs for $\dot{m}_{13} > 0$ as shown in Figure 5. To calculate the entrainment ($\dot{m}_{43}$ in this example), once again we use a plume description, but with an extended point source. The estimate for the point source extension is given by Cetegen *et al.* [22]. This virtual point source is chosen so that the flow at the door opening would correspond to a plume with the heating (with respect to the lower layer) given by

$$Q = c_p \left(T_1 - T_4 \right) \dot{m}_{13} \qquad (50)$$

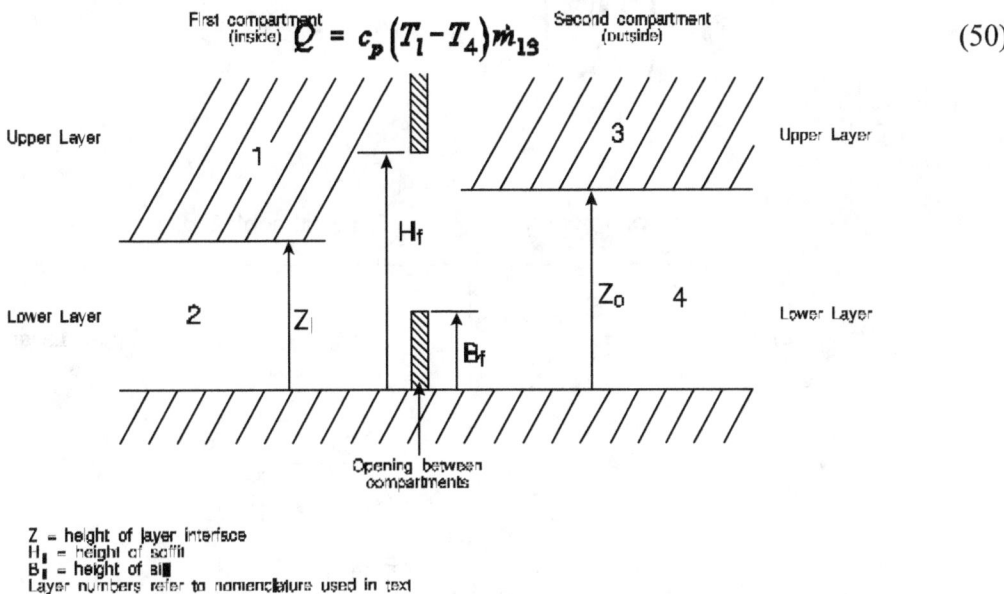

Z = height of layer interface
H_f = height of soffit
B_f = height of sill
Layer numbers refer to nomenclature used in text

Figure 4. Notation conventions for two-layer model in two compartments with a connecting vent.

The concept of the virtual source is that the enthalpy flux from the virtual point source should equal the actual enthalpy flux in the door jet at the point of exit from the vent using the same prescription. Thus the entrainment is calculated the same way as was done for a normal plume. The height, z_p, of the plume is

$$z_p = \frac{z_{13}}{Q_{eq}^{2/5}} + v_p \qquad (51)$$

where v_p, the virtual point source, is defined by inverting the entrainment process to yield

$$\nu_p = \left(\frac{90.9\dot{m}}{\dot{Q}_{eq}}\right)^{1.76} \qquad \text{if } 0.00 < \nu_p \le 0.08$$

$$\nu_p = \left(\frac{38.5\dot{m}}{\dot{Q}_{eq}}\right)^{1.001} \qquad \text{if } 0.08 < \nu_p \le 0.20 \qquad (52)$$

$$\nu_p = \left(\frac{8.10\dot{m}}{\dot{Q}_{eq}}\right)^{0.528} \qquad \text{if } 0.20 < \nu_p$$

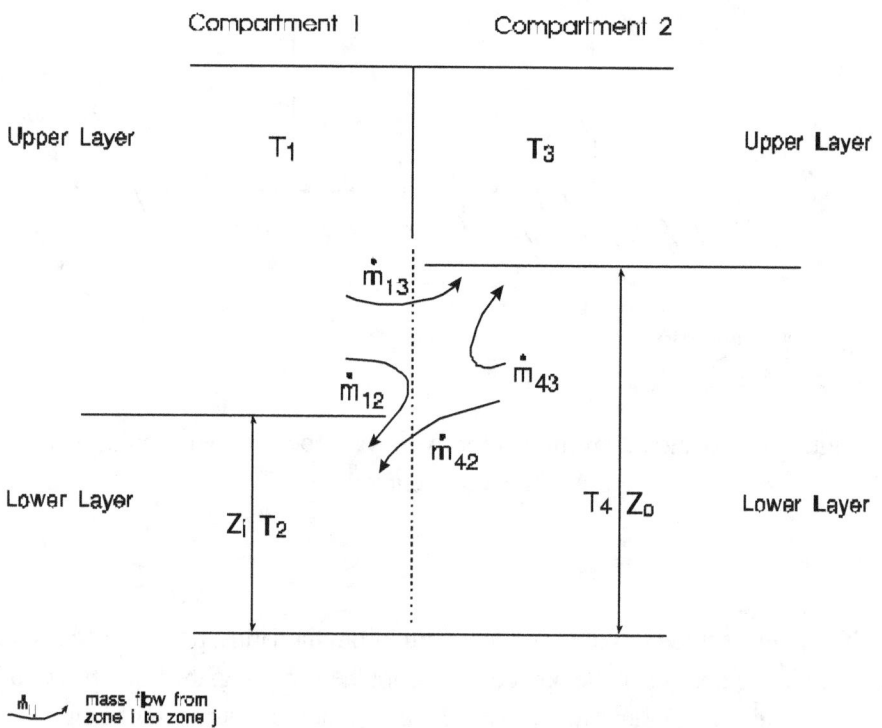

Figure 5. Flow patterns and layer numbering convention.

The units of this height, z_p and of ν_p, are not length, but rather the reduced notation of McCaffrey [21]. That is, the z_p defined here is the term $z/Q^{2/5}$ used earlier. Although outside of the normal range of validity of the plume model, a level of agreement with experiment is apparent (see section 5). Since a door jet forms a flat plume whereas a normal fire plume will be approximately circular, strong agreement is not expected.

30

The other type of mixing is much like an inverse plume and causes contamination of the lower layer. It occurs when there is flow of the type $\dot{m}_{42} > 0$. The shear flow causes vortex shedding into the lower layer and thus some of the particulates end up in the lower layer. The actual amount of mass or energy transferred is usually not large, but its effect can be large. For example, even minute amounts of carbon can change the radiative properties of the gas layer, from negligible to something finite. It changes the rate of radiation absorption by orders of magnitude and invalidates the simplification of an ambient temperature lower layer. This term is predicated on the Kelvin-Helmholz flow instability and requires shear flow between two separate fluids. The mixing is enhanced for greater density differences between the two layers. However, the amount of mixing has never been well characterized. Quintiere *et al.* discuss this phenomena for the case of crib fires in a single room, but their correlation does not yield good agreement with experimental data in the general case [28]. In the CFAST model, it is assumed that the incoming cold plume behaves like the inverse of the usual door jet between adjacent hot layers; thus we have a descending plume. It is possible that the entrainment is overestimated in this case, since buoyancy, which is the driving force, is not nearly as strong as for the usually upright plume.

3.3.2 <u>Vertical Flow Through Horizontal Vents</u>

Flow through a ceiling or floor vent can be somewhat more complicated than through door or window vents. The simplest form is uni-directional flow, driven solely by a pressure difference. This is analogous to flow in the horizontal direction driven by a piston effect of expanding gases. Once again, it can be calculated based on the Bernoulli equation, and presents little difficulty. However, in general we must deal with more complex situations that must be modeled in order to have a proper understanding of smoke movement. The first is an occurrence of puffing. When a fire exists in a compartment in which there is only one hole in the ceiling, the fire will burn until the oxygen has been depleted, pushing gas out the hole. Eventually the fire will die down. At this point ambient air will rush back in, enable combustion to increase, and the process will be repeated. Combustion is thus tightly coupled to the flow. The other case is exchange flow which occurs when the fluid configuration across the vent is unstable (such as a hotter gas layer underneath a cooler gas layer). Both of these pressure regimes require a calculation of the onset of the flow reversal mechanism.

Normally a non-zero cross vent pressure difference tends to drive unidirectional flow from the higher to the lower pressure side. An unstable fluid density configuration occurs when the pressure alone would dictate stable stratification, but the fluid densities are reversed. That is, the hotter gas is underneath the cooler gas. Flow induced by such an unstable fluid density configuration tends to lead to bi-directional flow, with the fluid in the lower compartment rising into the upper compartment. This situation might arise in a real fire if the room of origin suddenly had a hole punched in the ceiling. We make no pretense of being able to do this instability calculation analytically. We use Cooper's algorithm [29] for computing mass flow through ceiling and floor vents. It is based on correlations to model the unsteady component of the flow. What is surprising is that we can find a correlation at all for such a complex phenomenon. There are two components to the flow. The first is a net flow dictated by a pressure difference. The second is an exchange flow based on the relative densities of the gases. The overall flow is given by [29]

$$\dot{m} = C f(\gamma, \varepsilon) \left(\frac{\delta P}{\bar{P}} \right)^{1/2} A_v \qquad (53)$$

where $\gamma = c_p/c_v$ is the ratio of specific heats and

$$C = 0.68 + 0.17\varepsilon, \qquad (54)$$

$$\varepsilon = \frac{\delta P}{P}, \qquad (55)$$

and f is a weak function of both γ and ε. In the situation where we have an instability, we use Cooper's correlations. The algorithm for this exchange flow is given by

$$\dot{m}_{ex} = 0.1 \left(\frac{g \, \delta\rho \, A_v^{5/2}}{\rho_{av}} \right) \left(1.0 - \frac{2 A_v^2 \, \delta\rho}{S^2 \, g \, \delta\rho \, D^5} \right) \qquad (56)$$

where

$$D = 2 \sqrt{\frac{A_v}{\pi}} \qquad (57)$$

and S is 0.754 or 0.942 for round or square openings, respectively.

3.3.3 Forced Flow

Fan-duct systems are commonly used in buildings for heating, ventilation, air conditioning, pressurization, and exhaust. Figure 6(a) shows smoke management by an exhaust fan at the top of an atrium, and Figure 6(b) illustrates a kitchen exhaust. Cross ventilation, shown in Figure 6(c), is occasionally used without heating or cooling. Generally systems that maintain comfort conditions have either one or two fans. Residences often have a systems with a single fan as shown in Figure 6 and 7(a). In this system return air from the living quarters is drawn in at one location, flows through filter, fan and coils, and is distributed back to the residence. This system does not have the capability of providing fresh outside air. These systems are intended for applications where there is sufficient natural air leakage through cracks in walls and around windows and doors for odor control. Further information about these

32

systems is presented in Klote and Milke [30] and the American Society of Heating, Refrigerating and Air Conditioning Engineers [31].

The model for mechanical ventilation used in CFAST is based on the theory of networks and is based on the model developed by Klote [32]. This is a simplified form of Kirchoff's law which says that flow into a node must be balanced by flow out of the node. Adapting Ohm's law,

$$voltage = current \times resistance,$$

to HVAC flow, we have

$$pressure\ change = mass\ flow \times resistance$$

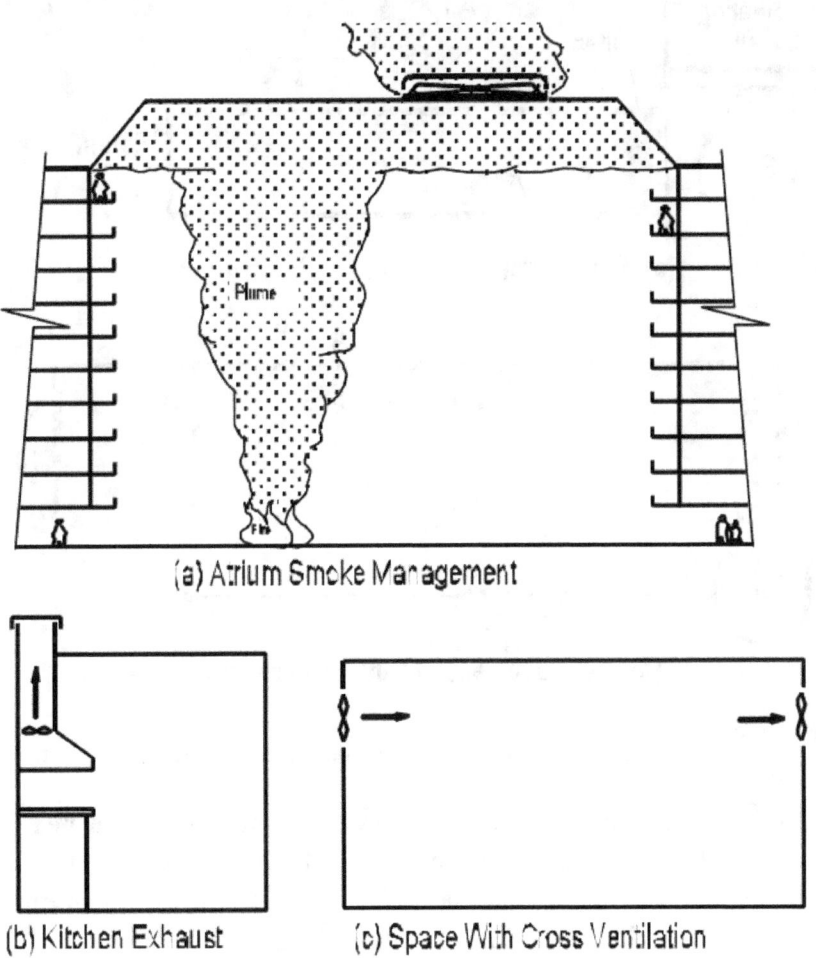

(a) Atrium Smoke Management

(b) Kitchen Exhaust   (c) Space With Cross Ventilation

Figure 6. Some simple fan-duct systems.

which can then be written equivalently

$$\text{mass flow} = \text{conductance} \times (\text{pressure drop across a resistance})^{1/2}.$$

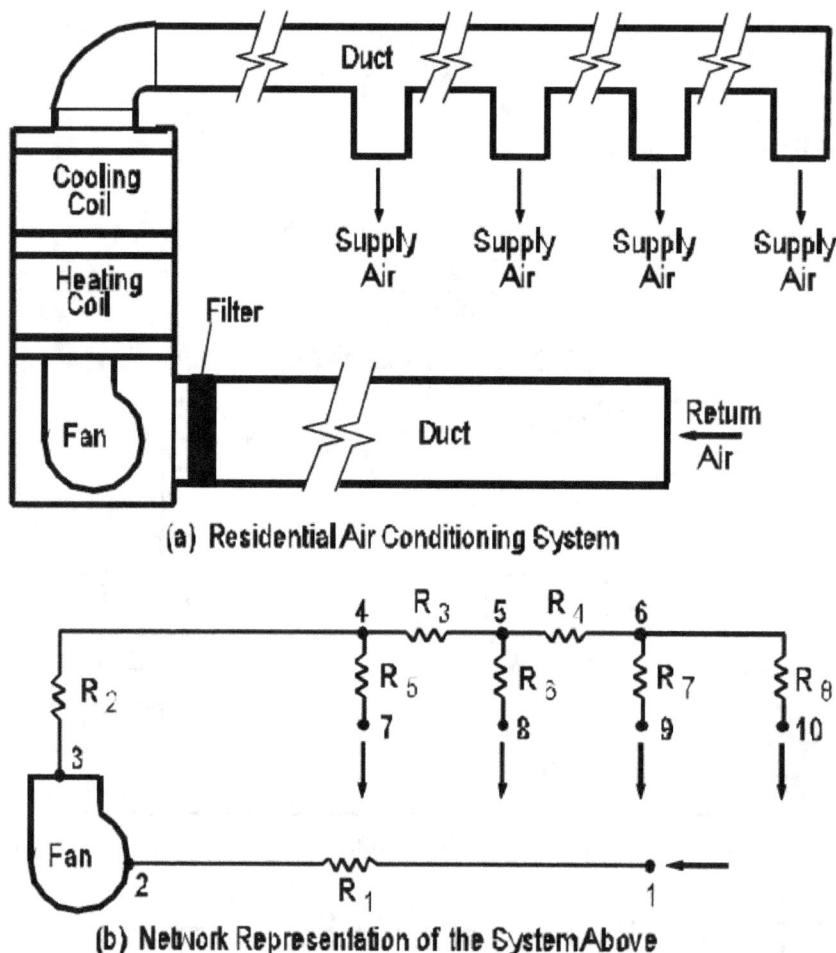

(a) Residential Air Conditioning System

(b) Network Representation of the System Above

Figure 7. Network representation of a residential system

For each node, this flow must sum to zero. There are several assumptions which are made in computing this flow in ducts, fans, elbow, *etc*. First, we assume unidirectional flow. Given the usual size of ducts, and the nominal presence of fans, this is quite reasonable. Also, the particular implementation used here [32] does not allow for reverse flow in the fans. The difficulty lies in describing how a fan behaves in such a case.

Each fan-duct system is represented as a network of nodes, each at a specific temperature and pressure. The nodes may be connected by fans, ducts, fittings and other components. Except for fans,

34

air flows through these components from nodes of higher pressure to nodes of lower pressure. For example, the residential system illustrated in 7(a) is represented in 7(b) as a network of a fan, eight resistances and ten nodes. These resistances incorporate all the resistance to flow between nodes. For instance, the equivalent resistance, R_1, between nodes 1 and 2 accounts for resistances of the inlet, duct, filter and connection to the fan.

Given that we can describe mass flow in terms of pressure differences and conductance, the conservation equation for each node is

$$f_i(P_1, P_2, \ldots) = \sum_j \dot{m}_{ij} = 0. \tag{58}$$

The index "j" is a summation over connections to a node, and there is an equation "i" for each node. The remaining problem is to specify the boundary conditions. At each connection to a compartment, the pressure is specified. Then, given that flow at each connection is unidirectional (at a given instant of time, the flow is either all into or all out of a given connection), the mass and enthalpy flow into or out of a room can be calculated explicitly. Thus we end up with a set of equations of the form

$$f_1(P_1, P_2, \ldots) = 0$$

$$\cdot$$

$$f_i(P_1, P_2, \ldots) = 0 \tag{59}$$

$$\cdot$$

$$\cdot$$

$$f_n(P_1, P_2, \ldots) = 0.$$

This is an algebraic set of equations that is solved simultaneously with the equations for flow in the compartments.

The equations describe the relationship between the pressure drop across a duct, the resistance of a duct, and the mass flow. The pressure can be changed by conditions in a compartment, or a fan in line in the duct system. Resistance arises from the finite size of ducts, roughness on surfaces, bends and joints. To carry the electrical analog a little further, fans act like constant voltage sources. The analogy breaks down in this case because the pressure (voltage) is proportional to the square of the velocity (current) rather than linearly related as in the electrical case. Since we are using the current form of the conservation equation to balance the system, the flow can be recast in terms of a conductance

$$\dot{m} = G\sqrt{\Delta P}. \tag{60}$$

The conductance can be expressed generally as

$$G = \sqrt{\frac{2\rho}{C_0}} \, A_0$$

(61)

where C_0 is the flow coefficient, and A_0 is the area of the inlet, outlet, duct, contraction or expansion joint, coil, damper, bend, filter, and so on. Their values for the most common of these items are tabulated in the ASHRAE Handbook [33].

The mechanical ventilation system is partitioned into one or more independent systems. Differential equations for species for each of these systems are derived by lumping all ducts in a system into one pseudo tank. This set of equations is then solved at each time step. Previously the mechanical ventilation computations in CFAST were performed as a side calculation using time splitting. This could cause problems since time-splitting methods require that the split phenomenon (the pressures and temperatures in this case) change slowly compared to other phenomenon such as room pressures, layer heights *etc.* The pressures at each internal node and the temperatures in each branch (duct, fan) are now determined explicitly by the solver, once again using conservation of mass and energy discussed in this section.

3.3.3.1 Ducts

Ducts are long pipes through which gases can flow. They have been studied much more extensively than other types of connections. For this reason, eq (61) can be put into a form which allows one to characterize the conductance in more detail, depending on the type of duct (e.g., oval, round, or square) and is given by

$$G = \sqrt{\frac{FL}{2\rho D_e A_0^2}} \, ,$$

(62)

where F is the friction factor, L and D_e are the length and effective diameter of the duct respectively. The temperature for each duct d is determined using the following differential equation:

accumulated heat = (heat in - heat out) - convective losses through duct walls

$$c_v \rho_d V_d \frac{dT_d}{dt} = c_p m_d \left(T_{in} - T_{out}\right) - h_d A_d \left(T_d - T_{amb}\right)$$

(63)

where c_v, c_p are the constant volume/pressure specific heats; V_d is the duct volume, ρ_d is the duct gas density, dT_d/dt is the time rate of change of the duct gas temperature, m_d is the mass flow rate, T_{in} and T_{out} are the gas temperatures going into and out of the duct, c_d, A_d are the convective heat transfer coefficient and surface area for duct d and T_{amb} is the ambient temperature. The first term on the right hand side of eq (63) represents the net gain of energy due to gas transported into or out of the duct. The second term represents heat transferred to the duct walls due to convection. In version 1.6, the loss coefficient is set to zero. We retain the form for future work. The differential and algebraic (DAE) solver used by CFAST solves eq (63) exactly as written. A normal ordinary differential equation solver would require that this equation be solved for dT/dt. By writing it this way, the duct volumes can be zero which is the case for fans.

3.3.3.2 Fans

This section provides background information about fan performance. For more information about fans, readers are referred to Jorgensen (1983) and ASHRAE (1992). Normal fan operating range is represented by the line segment AB in Figure 8. In this figure, Δp_f is the static pressure of the fan, and $\dot{V}_f$ is the volumetric flow of the fan. The point B represents a margin of safety selected by the fan manufacturer in order to avoid unstable flow.

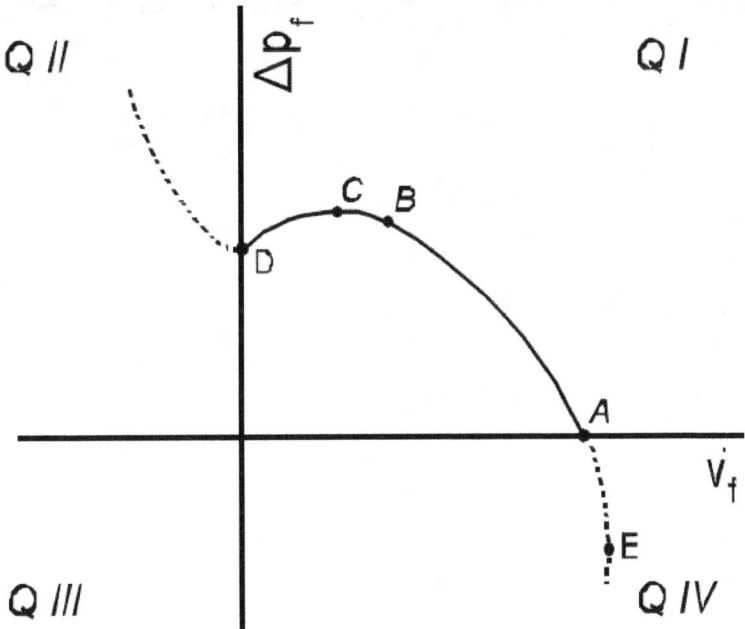

Figure 8. Typical fan performance at constant speed.

Fans operating in the positively sloping portion (*CD* of Figure 8) of the fan curve exhibit unstable behavior called surging or pulsing. Unstable flow consists of violent flow reversals accompanied by significant changes in pressure, power and noise. There is little information about how long a fan can operate in the unstable region before it is destroyed.

Backward flow through a fan occurs when the static pressure is greater than that at point *D*. This is also called second quadrant flow. Quadrant terminology is customarily used in description of fan performance. The horizontal axis and the vertical axis divide a plane into four quadrants which for convenience are labeled Q I, Q II, Q III and Q IV on Figure 8. Backward flow can be exhibited by all types of fans. The wind blowing into the outlet of a propeller fan can result in backflow, and pressures produced by fires could also produce backflow. Fourth quadrant flow is probably representative of all fans. As Δp_f becomes negative, the flow increases with decreasing Δp_f until a choking condition develops at point *E*.

It is common practice in the engineering community and fan industry to represent fan performance with Δp_f on the vertical axis and $\dot{V}_f$ on the horizontal axis. Probably the reason is that $\dot{V}_f$ can be thought of as a single valued function of Δp_f for flow in the first and second quadrants. Fan manufacturers generally supply flow-pressure data for the normal operating range, and they often supply data for the rest of the fan curve in the first quadrant. Specific data is not available for either second or fourth quadrant flow. No approach has been developed for simulation of unstable fan operation, and numerical modeling of unstable flow would be a complicated effort requiring research.

Numerical Approximation of Fan Performance: Figure 9 illustrates four approaches that can be used to approximate fan performance without simulation of unstable flow. For all of these approaches, the fan curve is used for the normal operating range of *AB*. Also for all of the approaches, flows above the normal operating range are approximated by a straight line tangent to the fan curve at point *A*. This results in fourth quadrant flow that is similar to the expected flow provided that Δp_f is not overly far below the horizontal axis. In Figure 9(a), flows below the normal range are approximated by a linear curve tangent to the fan curve at point *B*. This avoids simulation of unstable flows, but the approximated flow is higher than expected in the first quadrant and lower than expected for much of the fourth quadrant.

The approach of Figure 9(b) reduces the approximated flow in the first quadrant. In this approach, the fan curve is also used for the range *BF*, and flows above the normal operating range are approximated by a straight line tangent to the fan curve at point *F*. To increase the flow in the second quadrant,

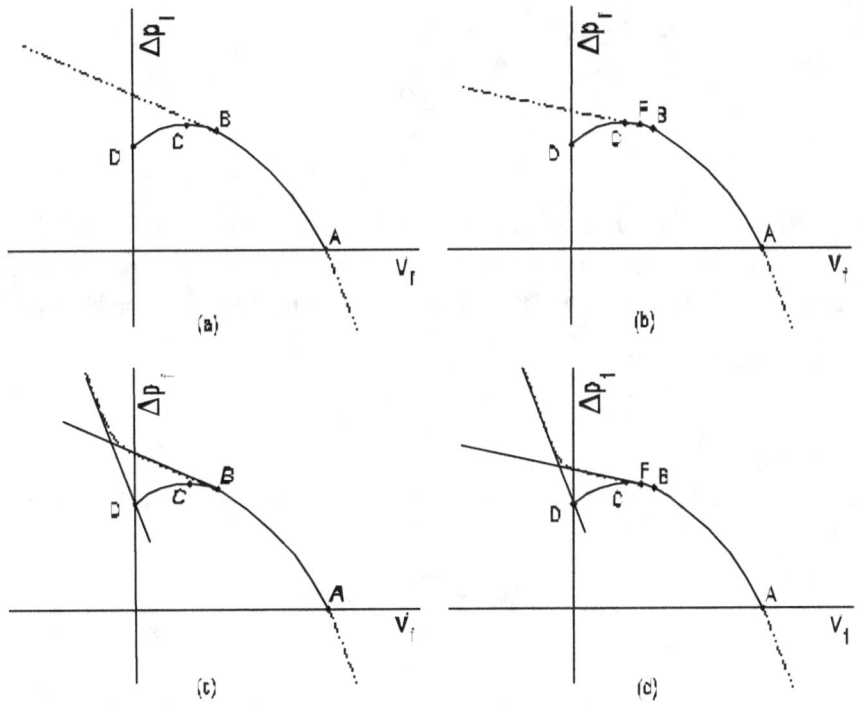

Figure 9. Some approaches to approximation of fan performance
for computer simulation

Figure 9(c) uses a line passing through point *D* with the slope of the fan curve at point *A*. Both of the modification of Figure 9(b) and Figure 9(c) are combined in the approach of Figure 9(d).

39

Fan manufacturer data is routinely either in tabular or graphical form. As indicated by Jorgensen [34], the use of a polynomial form of fan curve is common within the industry.

$$\dot{V}_f = B_1 + B_2 \Delta p_f + B_3 (\Delta p_f)^2 + \cdots + B_n (\Delta p_f)^{n-1} \tag{64}$$

The units for $\dot{V}_f$ and Δp_f in CFAST are m³/s and Pa respectively. Therefor the units for the coefficients B_i (i=1, ... , 5) are m³/(s Pa^{i-1}). The coefficients can be entered as data or calculated by least squares regression from flow and pressure data. For constant volumetric flow applications, the only non-zero coefficient in eq (64) is B_1 ($n = 1$). For incompressible fluids, eq (64) is independent of temperature and pressure. For fan data at 20 °C, compressibility effects amount to an error of about 6 % at a temperature of 200 °C.

3.3.3.3 Effective Resistance

The resistance, R, of a flow element can be defined as

$$R = \frac{\sqrt{\Delta p}}{\dot{m}} \tag{65}$$

where Δp is the pressure loss through the element corresponding to a mass flow rate, $\dot{m}$. The effective resistance between two nodes is always positive, however, sometimes one of the resistances between nodes can be negative as will be explained later. To account for this, $R = K^{\frac{1}{2}}$ can be substituted into eq (65) to give

$$\Delta p = K \dot{m}^2 \tag{66}$$

The total pressure loss, Δp_t, from one node to the next is the sum of the losses, Δp_i, through each flow element, i, between the nodes.

$$\Delta p_t = \sum_i \Delta p_i \tag{67}$$

The effective value, K_e, relates the total pressure loss to the mass flow rate as $\Delta p_t = K_e \dot{m}^2$, and K_i relates the pressure loss through element i as $\Delta p_i = K_i \dot{m}^2$. These pressure losses can be substituted into eq (67), and canceling like terms yields

$$K_e = \sum_i K_i \tag{68}$$

Values of K_i can be calculated for each element using equations developed later, and K_e can be calculated by eq (68).

Resistance of Ducts: For a straight section of duct with constant cross sectional area, the Bernoulli equation incorporating pressure loss, Δp_{fr}, due to friction is commonly written

$$p_1 - p_2 = \Delta p_{fr} + \rho g(Z_1 - Z_2) \tag{69}$$

where the subscripts 1 and 2 refer to the duct inlet and outlet respectively, p is pressure, Z is elevation, g is the acceleration due to gravity, and ρ is the density of the gas. The pressure loss due to friction is expressed by the Darcy equation in most elementary treatments of flow in pipes and ducts [35], [36], [37].

$$\Delta p_{fr} = f \frac{L}{D_e} \frac{\rho U^2}{2} \tag{70}$$

where f is the friction factor, L is the duct length, D_e is the effective diameter of the duct and U is the average velocity in the duct ($\dot{m} = \rho U A$ where A is the cross-sectional area of the duct). For a circular duct, the effective diameter is the duct diameter. For rectangular duct, Huebscher [38] developed the relationship

$$D_e = 1.30 \frac{(ab)^{0.625}}{(a + b)^{0.250}} \tag{71}$$

where a is the length of one side of the duct, and b is the length of the adjacent side. For flat oval duct, Heyt and Diaz [39] developed the relationship

$$D_e = \frac{1.55 A^{0.625}}{P^{0.200}} \tag{72}$$

where A and P are the cross-sectional area and the perimeter of the flat oval duct. The area of a flat oval duct is

$$A = (\pi b^2/4) + b(a - b) \tag{73}$$

and the perimeter of a flat oval duct is

41

$$P = \pi b + 2(a - b) \tag{74}$$

where a is the major dimension of the flat oval duct, and b is the minor dimension of the duct. Combining eqs 65 and 69 results in

$$\dot{m}_{ij} = \frac{1}{R}\sqrt{P_j - P_i + \rho g(Z_j - Z_i)} \tag{75}$$

Combining eqs 66 and 70 results in

$$K = \frac{fL}{2D_e \rho A^2} \tag{76}$$

where A is the cross sectional area of the duct. Colebrook developed the following equation for the friction factor [40].

$$\frac{1}{\sqrt{f}} = -2Log_{10}\left(\frac{\varepsilon}{3.7D_e} + \frac{2.51}{R_e\sqrt{f}}\right) \tag{77}$$

where R_e is the Reynolds number (UD_e/υ where υ is the kinematic viscosity) and ε is the roughness of the inside surface of the duct. Data on roughness of duct materials are listed in Table 3. A graphical presentation of the Colebrook equation developed by Moody [41] was used for decades to calculate friction factors. However, today it is practical to solve the Colebrook equation with computers.

Table 3. Absolute roughness values for common duct materials

Duct Material	Roughness Category	Absolute Roughness, ε	
		mm	ft
Uncoated Carbon Steel, Clean. PVC Plastic Pipe. Aluminum.	Smooth	0.03	0.0001
Galvanized Steel, Longitudinal Seams, 1200 mm Joints. Galvanized Steel, Continuously Rolled, Spiral Seams, 3000 mm Joints. Galvanized Steel, Spiral Seam with 1, 2 and 3 Ribs, 3600 mm Joints.	Medium Smooth	0.09	0.0003
Galvanized Steel, Longitudinal Seams, 760 mm Joints.	Average	0.15	0.0005
Fibrous Glass Duct, Rigid. Fibrous Glass Duct Liner, Air Side With Facing Material.	Medium Rough	0.9	0.003
Fibrous Glass Duct Liner, Air Side Spray Coated. Flexible Duct, Metallic. Flexible Duct, All Types of Fabric and Wire. Concrete.	Rough	3.0	0.01

Local Loss Resistances: The pressure loss, Δp, through many other elements can be expressed as

$$\Delta p = C_o \frac{\rho U_o^2}{2} \qquad (78)$$

where U_o is the average velocity at cross section o within the element, and C_o is a local loss coefficient. This equation is commonly used for inlets, outlets, duct contractions and expansions, heating and cooling coils, dampers, bends and many filters. For a large number of these elements, values of C_o have been empirically determined and are tabulated frequently as functions of geometry in handbooks [10 - 12]. Manufacturers literature also contains some values of C_o. The value of K for these resistances is

$$K = \frac{C_o}{2\rho A_o^2}$$

where A_o is the area at cross section o.

3.4 Corridor Flow

A standard assumption in zone fire modeling is that once hot smoke enters a compartment, a well defined upper layer forms instantly throughout the compartment. This assumption breaks down in large compartments and long corridors due to the time required to fill these spaces. A simple procedure is described for accounting for the formation delay of an upper layer in a long corridor by using correlations developed from numerical experiments generated with the NIST fire model FDS (Fire Dynamics Simulator) [42]. FDS is a computational fluid dynamics model capable of simulating fire flow velocities and temperatures with high resolution. Two parameters related to corridor flow are then estimated, the time required for a ceiling jet to travel in a corridor and the temperature distribution down the corridor. These estimates are then used in CFAST by delaying flow into compartments connected to corridors until the ceiling jet has passed these compartments.

IFS is used to estimate ceiling jet characteristics by running a number of cases for various inlet layer depths and temperatures. The vent flow algorithm in CFAST then uses this information to compute mass and enthalpy flow between the corridor and adjacent compartments. This is accomplished by presenting the vent algorithm with a one layer environment (the lower layer) before the ceiling jet reaches the vent and a two layer environment afterwards. Estimated ceiling jet temperatures and depths are used to define upper layer properties.

The problem is to estimate the ceiling jet temperature and depth as a function of time until it reaches the end of the corridor. The approach used here is to run a field model as a pre-processing step and to summarize the results as correlations describing the ceiling jet's temperatures and velocities. An outline of this process is given by

1.   Model corridor flow for a range of inlet ceiling jet temperatures and depths. Inlet velocities are derived from the inlet temperatures and depths.

2.   For each model run calculate average ceiling jet temperature and velocity as a function of distance down the corridor.

3.   Correlate the temperature and velocity distribution down the hall.

The zone fire model then uses these correlations to estimate conditions in the corridor.
An outline of the steps involved is given by

1.   Estimate the inlet temperature, depth and velocity of the ceiling jet. If the corridor is the fire room then use a standard correlation. If the source of the ceiling jet is another room then calculate the inlet ceiling jet flow using Bernoulli's law for the vent connecting the source room and the corridor.

44

2.   Use correlations in 3. above to estimate the ceiling jet arrival time at each vent.

3.   For each vent in the corridor use lower layer properties to compute vent flow before the ceiling jet arrives at the vent and lower/upper layer properties afterwards.

3.4.1 Assumptions

The assumptions made in order to develop the correlations are:

- The time scale of interest is the time required for a ceiling jet to traverse the length of the corridor. For example, for a 100 m corridor with 1 m/s flow, the characteristic time period would be 100 s.

- Cooling of the ceiling jet due to mixing with adjacent cool air is large compared to cooling due to heat loss to walls. Equivalently, we assume that walls are adiabatic. This assumption is conservative. An adiabatic corridor model predicts more severe conditions downstream in a corridor than a model that accounts for heat transfer to walls, since cooler ceiling jets travel slower and not as far.

- We do not account for the fact that ceiling jets that are sufficiently cooled will stagnate. Similar to the previous assumption, this assumption is conservative and results in over predictions of conditions in compartments connected to corridors (since the model predicts that a ceiling jet may arrive at a compartment when in fact it may have stagnated before reaching it).

- Ceiling jet flow is buoyancy driven and behaves like a gravity current. The inlet velocity of the ceiling jet is related to its temperature and depth.

- Ceiling jet flow lost to compartments adjacent to the corridor is not considered when estimating ceiling jet temperatures and depths. Similarly, a ceiling jet in a corridor is assumed to have only one source.

- The temperature and velocity at the corridor inlet is constant in time.

- The corridor height and width do not effect a ceiling jet's characteristics. Two ceiling jets with the same inlet temperature, depth and velocity behave the same when flowing in corridors with different widths or heights as long as the ratio of inlet widths to corridor width are equal.

45

- Flow entering the corridor enters at or near the ceiling. The inlet ceiling jet velocity is reduced from the vent inlet velocity by a factor of w_{vent}/w_{room} where w_{vent} and w_{room} are the width of the vent and room, respectively.

3.4.2 Corridor Jet Flow Characteristics

Ceiling jet flow in a corridor can be characterized as a one dimensional gravity current. To a first approximation, the velocity of the current depends on the difference between the density of the gas located at the leading edge of the current and the gas in the adjacent ambient air. The velocity also depends on the depth of the current below the ceiling. A simple formula for the gravity current velocity may be derived by equating the potential energy of the current, $mgd_0/2$, measured at the half-height $d_0/2$ with its kinetic energy, $mU^2/2$ to obtain

$$U = \sqrt{gd_0}$$

where m is mass, g is the acceleration of gravity, d_0 is the height of the gravity current and U is the velocity. When the density difference, between the current and the ambient fluid is small, the velocity U is proportional to $\sqrt{gd_0 \Delta \rho / \rho_{cj}} = \sqrt{gd_0 \Delta T / T_{amb}}$ where ρ_{amb}, T_{amb} are the ambient density and temperature and ρ_{cj}, T_{cj} are the density and temperature of the ceiling jet and $\Delta T = T_{cj} - T_{amb}$ is the temperature difference. Here use has been made of the ideal gas law, $\rho_{amb} T_{amb} \approx \rho_{cj} T_{cj}$ This can be shown using terms defined in Figure 10 by using an integrated form of Bernoulli's law noting that the pressure drop at the bottom of the ceiling jet is $P_b = 0$, the pressure drop at the top is $P_t = gd_0(\rho_{cj} - \rho_{amb})$ and using a vent coefficient c_{vent} of 0.74, to obtain

$$
\begin{aligned}
U_0 &= c_{vent}\frac{\sqrt{8}}{3}\frac{1}{\sqrt{\rho_{cj}}}\frac{P_t + \sqrt{P_t P_b} + P_b}{\sqrt{P_t} + \sqrt{P_b}} \\
&= c_{vent}\frac{\sqrt{8}}{3}\sqrt{P_t/\rho_{cj}} \\
&= c_{vent}\frac{\sqrt{8}}{3}\sqrt{gd_0\frac{\rho_{amb} - \rho_{cj}}{\rho_{cj}}} \\
&= 0.7\sqrt{gd_0\frac{\Delta T}{T_{amb}}}
\end{aligned}
$$

(81)

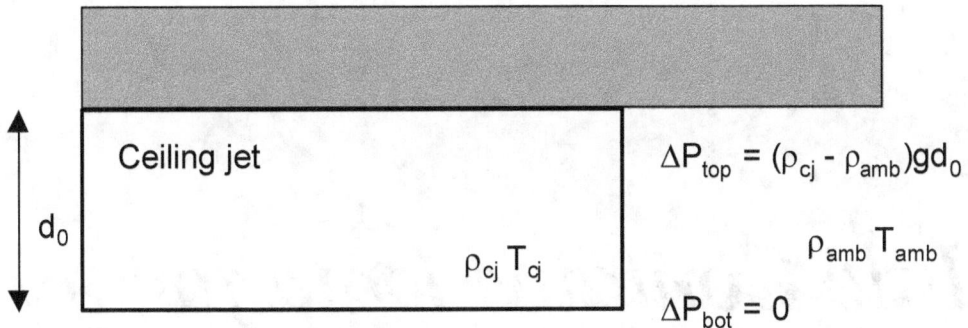

Figure 10 Schematic of a gravity current defining terms used to estimate its inlet velocity

Formulas of the form of the above equation lead one to conclude that a ceiling jet's characteristics in a corridor depend on its depth, d_0, and relative temperature difference, $\Delta T/T_{amb}$. Therefore, as the jet cools, it slows down. If no heat transfer occurs between the ceiling jet and the surrounding walls, then the only mechanism for cooling is mixing with surrounding cool air.

Twenty numerical experiments were performed using the field model IFS [42] in order to better understand the effects of the inlet ceiling jet temperature and depth on ceiling jet characteristics downstream in a corridor. These cases were run with five different inlet depths and four different inlet temperatures. The inlet ceiling jet temperature rise, ΔT_0, and depth, d_0, were used to define an inlet velocity, U_0 using eq(81). The inlet ceiling jet depths, d_0, used in the parameter study are 0.15 m, 0.30 m, 0.45 m, 0.60 m and 0.75 m. The inlet ceiling jet temperature rises, T_0, used in the parameter study are 100 °C, 200 °C, 300 °C and 400 °C.

3.4.3 Correlations

Ceiling jet functions were plotted as a function of distance down a corridor for each of the twenty test cases. These results are shown in Figure 11. Note that all but the 0.15 m ceiling jet data lies on essentially the same line.

The best fit line is given in the form of

$$\log\frac{\Delta T}{\Delta T_0} = a + bx.$$

File Contains Data for PostScript Printers Only

Figure 11 Common log of relative temperature excess downstream in a corridor using an adiabatic temperature boundary condition for several inlet depths and inlet temperature rises.

This is equivalent to

$$\frac{\Delta T}{\Delta T_0} = C_1 10^{bx} = C_1 \left(\frac{1}{2}\right)^{x/h_{1/2}}$$

where $C_1 = 10^a$ and $h_{1/2} = -\log(2)/b$. The parameter $h_{1/2}$ has a physical interpretation. It is the distance down the corridor where the temperature rise T, falls off to 50 per cent of its original value or equivalently, $T(x + h_{1/2}) = T(x)/2$.

The half-distance, $h_{1/2}$, can be approximated by $h_{1/2} = \log(2)/0.018 = 16.7$ m where $b = -0.018$ is given in Figure 11. Similarly, the coefficients C_1 is approximated by $C_1 = 10^a = 10^{-0.003} \approx 1$ where a is also given in Figure 11. Therefore the temperature rise, ΔT, may be approximated by

48

$$\Delta T = \Delta T_0 \left(\frac{1}{2} \right)^{\frac{x}{16.7}}$$

The numerical experiments with IFS demonstrated that for the cases simulated, ceiling jet characteristics depend on the relative inlet temperature rise and not the inlet depth. Flow in long corridors (greater than 10 m) need to be better characterized due to the flow stagnation which may occur because of the ceiling jet's temperature decay.

3.5 Heat Transfer

This section discusses radiation, convection and conduction, the three mechanisms by which heat is transferred between the gas layers and the enclosing compartment walls. This section also discusses heat transfer algorithms for calculating target temperatures.

3.5.1 Radiation

Objects such as walls, gases and fires radiate as well as absorb radiation. Each object has its own properties, such as temperature and emissivity. As we are solving the enthalpy equation for the gas temperature, the primary focus is in finding out how much enthalpy is gained or lost by the gas layers due to radiation. To calculate the radiation absorbed in a zone, a heat balance must be done which includes all surfaces which radiate to and absorb radiation from a zone. The form of the terms which contribute heat to an absorbing layer are the same for all layers. Essentially we assume that all zones in these models are similar so we can discuss them in terms of a general layer contribution. For this calculation to be done in a time commensurate with the other sources, some approximations are necessary.

Radiation can leave a layer by going to another layer, by going to the walls, by exiting through a vent, by heating an object, or by changing the pyrolysis rate of the fuel source. Similarly, a layer can be heated by absorption of radiation from these surfaces and objects as well as from the fire itself. The formalism which we employ for the geometry and view factor calculation is that of Siegel and Howell [44]. Although the radiation could be done with a great deal of generality, we have assumed that the zones and surfaces radiate and absorb like a grey body.

Radiation is an important mechanism for heat exchange in compartments subject to fires. It is important in the present application because it can affect the temperature distribution within a compartment, and thus the buoyancy forces. In the present implementation the fire is assumed to be a point source; it is assumed that plumes do not radiate. We use a simplified geometrical equivalent of the compartment in order to calculate the radiative transfer between the ceiling, floor and layer(s). The original paper which described FAST pointed out that there was an inconsistency in the interaction between the two wall radiation and the four wall conduction algorithms used to transfer heat between the gas layers and the walls. A four wall radiation heat transfer algorithm fixes this problem. A radiative heat transfer calculation could easily dominate the computation in any fire model. This is because radiation exchange is a global phenomena. Each portion of an enclosure interacts radiatively with every other portion that it "sees." Therefore, it is important to construct algorithms for radiative heat transfer that are both accurate and efficient [43].

This is a "next step" algorithm for computing radiative heat transfer between the bounding surfaces of a compartment containing upper and lower layer gases and point source fires. The two-wall radiation model used has been enhanced to treat lower layer heating and to treat radiative heat exchange with the upper and lower walls independently of the floor and ceiling. We refer to this as the four wall model.

The original radiation algorithm used the extended floor and ceiling concept for computing radiative heat exchange. For the purposes of this calculation, the room is assumed to consist of two wall segments: an extended ceiling and an extended floor. The extended ceiling consisted of the ceiling plus the upper wall segments. Similarly, the extended floor consisted of the floor plus the lower wall segments. The upper layer was modeled as a sphere equal in volume to the volume of the upper layer. Radiative heat transfer to and from the lower layer was ignored. This algorithm is inconsistent with the way heat conduction is handled, since we solve up to four heat conduction problems for each room: the ceiling, the upper wall, the lower wall and the floor. The purpose of the new radiation algorithm then is to enhance the radiative module to allow the ceiling, the upper wall segments, the lower wall segments and the floor to transfer radiant heat independently and consistently.

The four wall algorithm for computing radiative heat exchange is based upon the equations developed in Siegel and Howell [44] which in turn is based on the work of Hottel [45]. Siegel and Howell model an enclosure with N wall segments and an interior gas. A radiation algorithm for a two layer zone fire model requires treatment of an enclosure with two uniform gases. Hottel and Cohen [46] developed a method where the enclosure is divided into a number of wall and gas volume elements. An energy balance is written for each element. Each balance includes interactions with all other elements. Treatment of the fire and the interaction of the fire and gas layers with the walls is based upon the work of Yamada and Cooper [47]. They model fires as point heat sources radiating uniformly in all directions and use the Lambert-Beer law to model the interaction between heat emitting elements (fires, walls, gas layers) and the gas layers. The original formulation is for an N-wall configuration. Although this approach would allow arbitrary specification of compartment surfaces (glass window walls, for example), the computational requirements are significant.

Even the more modest approach of a four wall configuration for computing radiative heat transfer is more sophisticated than was used previously. By implementing a four wall rather than an N wall model, significant algorithmic speed increases were achieved. This was done by exploiting the simpler structure and symmetry of the four wall problem.

The radiation exchange at the k'th surface is shown schematically in Figure 12. For each wall segment k from 1 to N, we must find a net heat flux, $\Delta q_k''$, such that

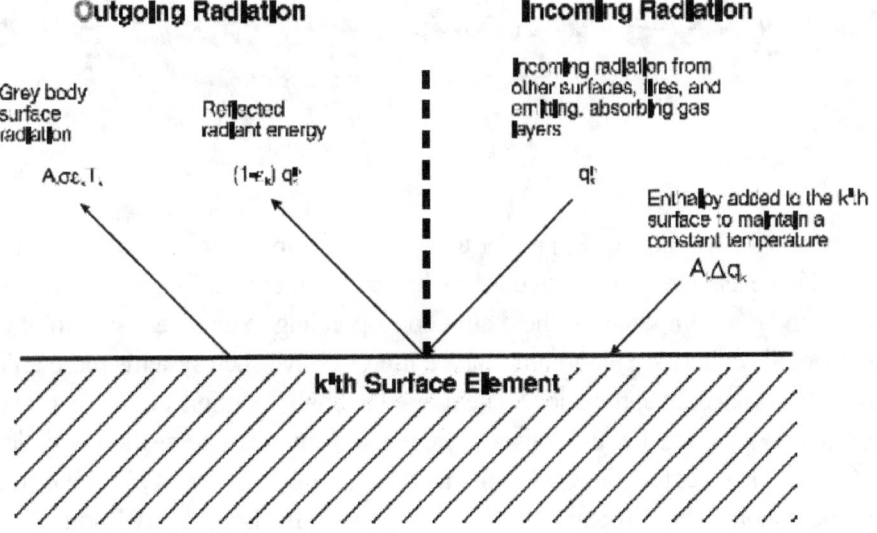

Figure 12. Radiation exchange in a two-zone fire model.

$$A_k \epsilon_k \sigma_k T_k^{\ 4} + (1-\epsilon_k) q_k^{in} = q_k^{in} + A_k \Delta q_k'' \quad (k=1,..N). \tag{85}$$

Radiation exchange at each wall segment has emitted, reflected, incoming and net radiation terms. Equation (85) then represents a system of linear equations that must be solved for $\Delta q''$ to determine the net fluxes given off by each surface. The setup and solution of this linear system is the bulk of the work required to implement the net radiation method of Siegel and Howell. Equation (86) derived by Siegel and Howell [44] and listed there as eqs 17 to 20, is called the net radiation equation,

$$\frac{\Delta q_k''}{\epsilon_k} - \sum_{j=1}^{N} \frac{1-\epsilon}{\epsilon_j} \Delta q_j'' F_{k-j} \tau_{j-k} = \sigma T_k^{\ 4} - \sum_{j=1}^{N} \sigma T_j^{\ 4} F_{k-j} \tau_{j-k} - \frac{c_k}{A_k}. \tag{86}$$

where σ is the Stefan-Boltzman constant, ϵ_k is the emissivity of the k'th wall segment, T_k is the temperature of the k'th wall segment, F_{k-j} a configuration factor, and τ is a transmissivity factor. This latter is the fraction of energy passing unimpeded through a gas along a path from surface j to k. The

51

parameters c_k represent the various sources of heat, namely the fire itself and the gas layers. In the form shown, the view factor of the k'th element is included in the parameter c_k.

The actual implementation uses a slightly modified form of eq (86), namely

$$\Delta \hat{q}_k'' - \sum_{j=1}^{N} (1-\epsilon_j) \Delta \hat{q}_j'' F_{k-j} \tau_{j-k} = \sigma T_k^4 - \sum_{j=1}^{N} \sigma T_j^4 F_{k-j} \tau_{j-k} - \frac{c_k}{A_k}, \text{ where} \quad (87)$$

$$\Delta q_k'' = \epsilon_k \Delta \hat{q}_k''. \quad (88)$$

There are two reasons for solving eq (87) rather than eq (86). First, since ϵ_k does not occur in the denominator, radiation exchange can be calculated when some of the wall segments have zero emissivity. Second and more importantly, the matrix corresponding to the linear system of eq () is diagonally dominant [43]. Iterative algorithms can be used to solve such systems more efficiently than direct methods such as Gaussian elimination. The more diagonally dominant a matrix (the closer the emissivities are to unity), the quicker the convergence when using iterative methods. Typical values of the emissivity for walls subject to a fire environment are in the range of $0.85 < \epsilon < 0.95$, so this is a reasonable approximation. The computation of, F_{k-j}, τ_{j-k} and c_k is discussed by Forney [43]. It is shown how it is possible to use the symmetries present in the four wall segment problem to minimize the number of direct configuration factor calculations required. In earlier versions of CFAST, the gas transmittance per unit length was assumed constant. In this new version, is calculated from the properties of the gas layers. Appendix A provides details of the gas transmittance calculation.

For rooms containing a fire, CFAST models the temperature of four wall segments independently. A two wall model for radiation exchange can break down when the temperatures of the ceiling and upper walls differ significantly. This typically happens in the room of fire origin when different wall materials are used as boundaries for the ceiling, walls and floor.

To demonstrate this consider the following example. To simplify the comparison between the two and four wall segment models, assume that the wall segments are black bodies (the emissivities of all wall segments are one) and the gas layers are transparent (the gas absorptivities are zero). This is legitimate since for this example we are only interested in comparing how a two wall and a four wall radiation algorithm transfer heat to the wall segments. Let the room dimensions be (4×4×4) m, the temperature of the floor and the lower and upper walls be 300 K. Let the ceiling temperature vary from 300 K to 600 K.

Figure 13 shows a plot of the heat flux to the ceiling and upper wall as a function of the ceiling temperature [43], [48]. The two wall model predicts that the extended ceiling (a surface formed by combining the ceiling and upper wall into one wall segment) cools, while the four wall model predicts

52

that the ceiling cools and the upper wall warms. The four-wall model moderates temperature differences that may exist between the ceiling and upper wall (or floor and lower wall) by allowing heat transfer to occur between the ceiling and upper wall. The two wall model is unable to predict heat transfer between the ceiling and the upper wall since it models them both as one wall segment.

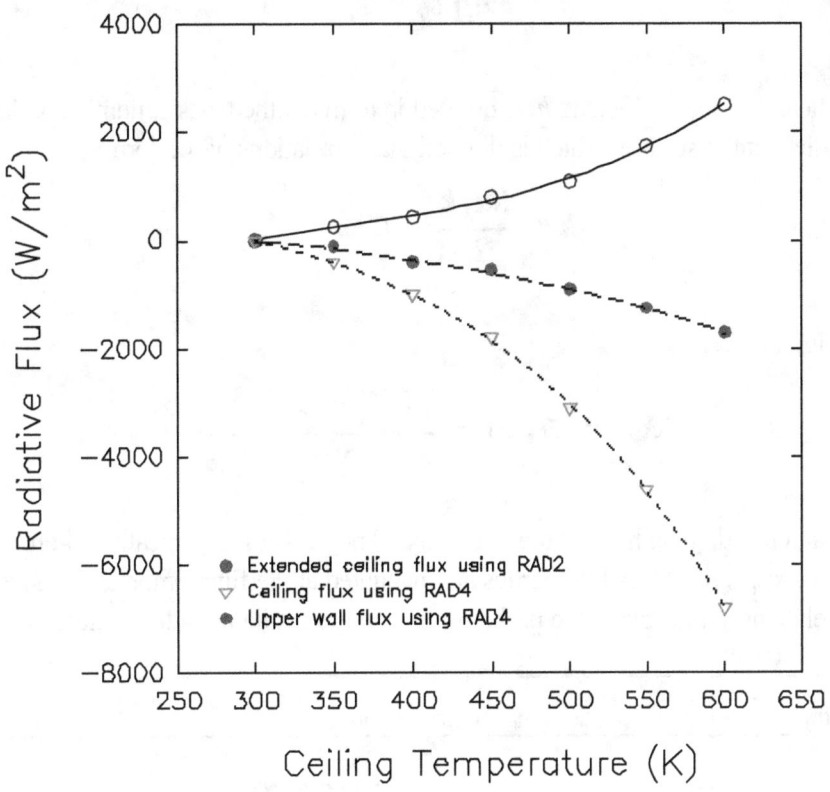

Figure 13. An example of two-wall and four-wall calculations for radiation exchange on a ceiling and wall surface.

3.5.2 Convection

Convection is one of the mechanisms by which the gas layers lose or gain energy to walls, objects or through openings. Conduction is a process which is intimately associated with convection; but as it does not show up directly as a term for heat gain or loss, it will be discussed separately. Convective heating describes the energy transfer between solids and gases. The enthalpy transfer associated with flow through openings was discussed in the section on flow through vents.

Convective heat flow is enthalpy transfer across a thin boundary layer. The thickness of this layer is determined by the temperature difference between the gas zone and the wall or object being heated [49]. In general, convective heat transfer, $\dot{q}$, is defined as

$$\dot{q} = hA_s\left(T_g - T_s\right).\tag{89}$$

The convective heat transfer coefficient, h, is defined in terms of the Nusselt number, a dimensionless temperature gradient at the surface, which is defined via correlations of the form

$$h = \frac{Nu_L k}{L} = C Ra_L^n \tag{90}$$

where the Rayleigh number,

$$Ra_L = Gr_L \, Pr = \frac{g\beta\left(T_s - T_g\right)L^3}{\nu\alpha} \tag{91}$$

is based on a characteristic length, L, of the geometry. The power n is typically 1/4 and 1/3 for laminar and turbulent flow, respectively. All properties are evaluated at the film temperature, $T_f \equiv (T_s + T_g)/2$. The typical correlations applicable to the problem at hand are available in the literature [50]:

Geometry	Correlation	Restrictions
Walls	$Nu_L = \left(0.825 + \dfrac{0.387 Ra_L^{1/6}}{\left(1 + (0.492/Pr)^{9/16}\right)^{8/27}}\right)^2$ $\approx 0.12\, Ra_L^{1/3}$	none
Ceilings and floors (hot surface up or cold surface down)	$Nu_L = 0.13\, Ra_L^{1/3}$	$2 \cdot 10^8 \leq Ra_L \leq 10^{11}$
Ceilings and floors (cold surface up or hot surface down)	$Nu_L = 0.16\, Ra_L^{1/3}$	$10^8 \leq Ra_L \leq 10^{10}$

The thermal diffusivity, α, and thermal conductivity, k, of air are defined as a function of the film temperature from data in reference [50].

$$\alpha = 1.0 \times 10^{-9} \; T_f^{7/4}$$

$$k = \left(\frac{0.0209 + 2.33 \times 10^{-5} T_f}{1 - 0.000267 T_f} \right) \tag{92}$$

Implementation: The algorithm is implemented as described above in the routine CONVEC in CFAST. The values of the correlation coefficients have changed in the new algorithm. In the old implementation, these were 0.21 and 0.012 for hot surface up and hot surface down respectively. In the new implementation, these are 0.16 and 0.13. The new values come from currently accepted engineering literature pertaining to this general area. As an example of the effect of the new algorithm, Figure 14 shows layer and wall temperatures for a single room test case with a relatively small 100 kW fire. Only minor differences are seen in the temperatures. For larger fires, the effect should be even less pronounced since convection will play a lesser role as the temperatures rise. A major advantage in the new algorithm should be a speed increase due to the closer correlation coefficients and smooth transition for cases where the surfaces are cooling – a point where the existing code tends to slow dramatically.

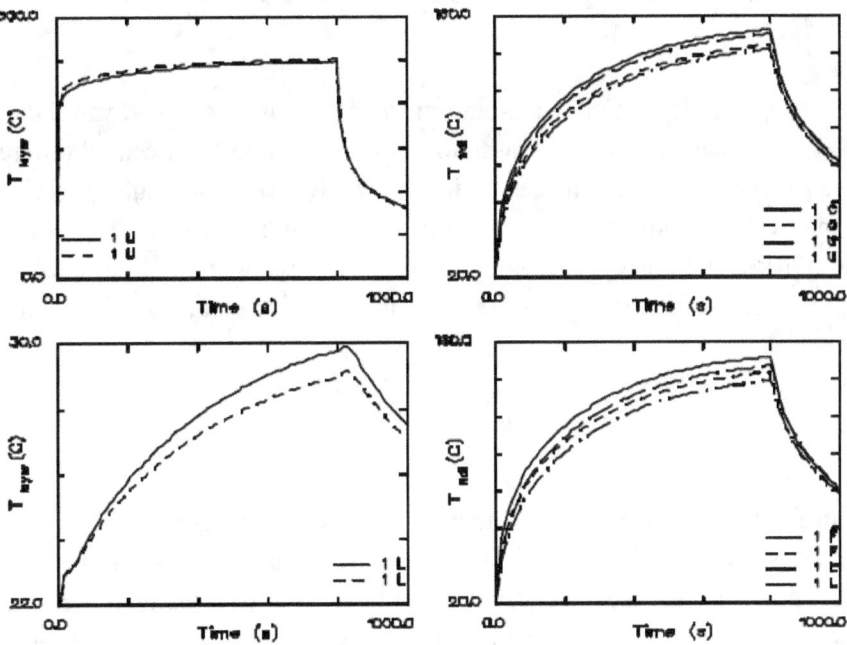

Figure 14. Effects on layer and wall temperatures of
modifications to the convection algorithm in CFAST.

3.5.3 Conduction

Heat loss or gain between a wall and a gas layer is due to convection or radiation not conduction. Conduction transfers heat within the wall. Therefore, source terms for conduction do not appear in the gas layer ordinary differential equations. However, convection and radiative heat transfer calculations provide the boundary conditions for the conduction algorithm discussed below.

The partial differential equation which governs the heat transfer in solids is

$$\frac{\partial T}{\partial t} = \frac{k}{\rho c} \nabla^2 T \tag{93}$$

It must be solved by a different technique than is used for the ordinary differential equations used to model gas layer quantities. We assume that the coefficients k, ρ and c are independent of temperature throughout the material. This may not be the case, especially for some materials such as gypsum for which the value of k may vary by a factor of two or more. However, to the accuracy that we know most of the thermal properties, it is a reasonable approximation. Procedures for solving 1-d heat conduction problems are well known. For finite difference methods such as backward difference (fully implicit), forward difference (fully explicit) or Crank-Nicolson, see [51]. For finite element methods see [52].

A finite difference approach [53] using a non-uniform spatial mesh is used to advance the wall temperature solution. The heat equation is discretized using a second order central difference for the spatial derivative and a backward differences for the time derivative. The resulting tri-diagonal system of equations is then solved to advance the temperature solution to time $t + \Delta T$. This process is repeated , using the work of Moss and Forney [53], until the heat flux striking the wall (calculated from the convection and radiation algorithms) is consistent with the wall temperature gradient at the surface via Fourier's law

$$q'' = -K\frac{dT}{dx}$$

where K is the thermal conductivity. This solution strategy requires a differential algebraic (DAE) solver that can simultaneously solve both differential (gas ODE's) and algebraic equations (Fourier's law). With this method, only one or two extra equations are required per wall segment (two if both the interior and exterior wall segment surface temperatures are computed). This solution strategy is more efficient than the method of lines since fewer equations need to be solved. Wall segment temperature profiles, however, still have to be stored so there is no decrease in storage requirements. Conduction is then coupled to the room conditions by temperatures supplied at the interior boundary by the differential equation solver. The exterior boundary condition types (constant flux, insulated, or constant temperature) are specified in the configuration of CFAST.

A non-uniform mesh scheme was chosen to allow breakpoints to cluster near the interior and exterior wall segment surfaces. This is where the temperature gradients are the steepest. A breakpoint x_b was defined by $x_b=MIN(x_p, W/2)$, where $x_p = 2(\alpha t_{final})^{\frac{1}{2}} erfc^{-1}(.05)$ and erfc^{-1} denotes the inverse of the complementary error function. The value x_p is the location in a semi-infinite wall where the temperature rise is 5 % after t_{final} seconds and is sometimes called the penetration depth. Eighty % of the breakpoints were placed on the interior side of x_b and the remaining 20 % were placed on the exterior side.

To illustrate the method, consider a one room case with one active wall. There will be four gas equations (pressure, upper layer volume, upper layer temperature, and lower layer temperature) and one wall temperature equation. Implementation of the gradient matching method requires that storage be allocated for the temperature profile at the previous time, t, and at the next time, $t + \delta t$. Given the profile at time t and values for the five unknowns at time $t + \delta t$ (initial guess by the solver), the temperature profile is advanced from time t to time $t + \delta t$. The temperature profile gradient at $x = 0$ is computed followed by the residuals for the five equations. The DAE solver adjusts the solution variables and the time step until the residuals for all the equations are below an error tolerance. Once the solver has completed the step, the array storing the temperature profile for the previous time is updated, and the DAE solver is ready to take its next step.

One limitation of our implementation of conduction is that it serves only as a loss term for enthalpy. Heat lost from a compartment by conduction is assumed to be lost to the outside ambient. In reality, compartments adjacent to the room which contains the fire can be heated, possibly catastrophically, by conducted energy not accounted for in the model. Although solving the conduction equations for this situation is not difficult, the geometrical specification is. For this reason, we have chosen to assume that the outside of a boundary is always the ambient. A means to connect compartments physically so that heat can be transported by conduction is under active study.

3.5.4 Inter-compartment Heat Transfer

Heat transfer between vertically connected compartments is modeled by merging the connected surfaces for the ceiling and floor compartments or for the connected horizontal compartments. A heat conduction problem is solved for the merged walls using a temperature boundary condition for both the near and far wall. As before, temperatures are determined by the DAE solver so that the heat flux striking the wall surface (both interior and exterior) is consistent with the temperature gradient at that surface. This option is implemented with the CFCON (for vertical heat transfer) and the HHEAT (for horizontal heat transfer) keywords.

For horizontal heat transfer between compartments, the connections can be between partial wall surfaces, expressed as a fraction of the wall surface. CFAST first estimates conduction fractions analogous to radiative configuration factors. For example, a conduction fraction between a rear wall in room 1 and a front wall room 2 is the heat flux fraction from the room 2 wall that strikes contributes to room 1's wall heat transfer. Alternatively, these fractions can be specified on the HHEAT keyword line. Once these fractions are determined, an average flux, q_{avg}, is calculated using

$$q_{avg} = \sum_{walls} F_{ij} q_{wall_j}$$

where F_{ij} is the fraction of flux from wall i that contributes to wall j, q_{wallj} is the flux striking wall j

3.5.5 Heating of Targets

The target calculation is similar to the heat conduction through boundaries. The net flux striking a target can be used as a boundary condition for an associated heat conduction problem in order to compute the surface temperature of the target. This temperature can then be used to estimate the conditions at the target, ie whether the target will ignite. Alternatively, if the target is assumed to be thin, then its temperature quickly rises to a level where the net heat flux striking the target is zero, ie to a steady state. The calculation is done using the concept of net heat flux, which literally implies a Kirchoff law for the radiation. While this should be obvious, it has not always been done this way in fire modeling.

The net heat flux, $\Delta''q_t$, striking a target t is given by

$$\Delta''q_t = q''_{rad}(in) + q''_{convec} - q''_{rad}(out) \tag{95}$$

where $q''_{rad}(in)$ is the incoming radiative flux, q''_{convec} is the convective flux and $q''_{rad}(out)$ is the outgoing radiative flux, Figure 15. The incoming radiative flux can be split into components from each fire, $q''_{f,t}$, each wall segment, $q''_{w,t}$ and each gas layer, $q''_{g,t}$. The incoming radiative flux, $q''_{rad}(in)$ is obtained by summing over all fires, wall segments, and gas layers to obtain

$$q_{rad}(in) = \sum_f q''_{f,t} + \sum_w q''_{w,t} + \sum_g q''_{g,t} \tag{96}$$

The outgoing radiative flux, $q''_{rad}(out)$, has contributions due to target emission or emissive power, $\epsilon_t \sigma T^4_t$, and a fraction, $1-\epsilon_t$, of the incoming radative flux that is reflected at the target surface. It is given by

$$q_{rad}^{out} = (1-\epsilon_t)q_{rad}^{in} + \epsilon_t \sigma T_t^4 \tag{97}$$

where ϵ_t is the emittance of the target and σ is the Stefan-Boltzman constant. Substituting eqs (96) and (97) into (95) we obtain

File Contains Data for PostScript Printers Only

Figure 15. Radiative and convective heat transfer at a target.

$$\Delta''q_t = \epsilon_t q_{rad}''(in) + q_{convec}'' - \epsilon_t \sigma T_t^4 = \epsilon_t \left(\sum_f q_{f,t}'' + \sum_w q_{w,t}'' + \sum_t q_{g,t}'' \right) + q_{convec}'' - \epsilon_t \sigma T_t^4 \tag{98}$$

The heat flux, $\Delta''q_t$, specified by eq(98) can be used in one of two ways: to estimate the surface temperature of the target or as a boundary condition to solve the heat conduction problem

$$\frac{\partial T}{\partial t}(x,t) = \frac{k}{\rho C}\frac{\partial^2 T}{\partial x^2}(x,t) \tag{99}$$

$$T(x,0) = T_0(x) \tag{100}$$

$$-k\frac{\partial T}{\partial x}(0,t) = \Delta''q_t \tag{101}$$

$$-k\frac{\partial T}{\partial x}(L,t) = 0 \tag{102}$$

for the target temperature profile T where k, ρ and C are the thermal conductivity, density and heat capacity of the target, L is the thickness of the target and T_0 is the target's initial temperature profile.

Alternatively, we can assume that the target temperature is always at steady state, *i.e.* that $\Delta''q_t = 0$. A temperature, T_t, can then be found using Newton's method that will satisfy $\Delta''q_t = 0$. To illustrate, suppose that the convective flux is given by $q''_{convec} = c(T_g-T_t)$ where c is a convective heat transfer coefficient and T_g is the gas temperature adjacent to the target. Then, eq(98) can be simplified to

$$f(T_t) = \epsilon_t\sigma T_t^4 - c\left(T_g - T_t\right) - \epsilon_t q''_{rad}(in) = -\Delta''q_t \tag{103}$$

We wish to find a temperature T_t satisfying $f(T_t) = -\Delta''q_t = 0$. The non-linear eq(103) can be solved using Newton's method. There are three steps necessary to complete this calculation: 1) calculate the heat transfer through the compartment; 2) calculate the heat flux to the target or object; and finally 3) compute the target temperature. In order to calculate the radiation heat transfer from fires, gas layers and wall surfaces to targets we must be able to calculate configuration factors, gas layer transmissivity and absorptance.

Configuration Factors: Figure 16 illustrates the definition of a configuration factor between two finite areas. A configuration factor between two finite areas 1 and 2 denoted F_{1-2} is the fraction of radiant energy given off by surface 1 that is intercepted by surface 2. The following terms are needed to define a configuration factor mathematically. Vectors n_1 and n_2 are unit vectors perpendicular to surfaces 1 and 2. s is a vector with origin on surface 1 and destination on surface 2. $S = \|s\|$ is the length of this vector. The angle between n_1 and s is θ_1. Similarly the angle between the vector n_2 and $-s$ is θ_2. The sign is reversed because the origin of s is on surface 1 not 2. The cosines of angles θ_1 and θ_2 are

$$\cos(\theta_1) = \frac{n_1 \cdot s}{|s|} \tag{104}$$

$$\cos(\theta_2) = -\frac{n_2 \cdot s}{|s|} \qquad (105)$$

File Contains Data for PostScript Printers Only

Figure 16. Setup for a configuration factor calculation between two arbitrarily oriented finite areas

The configuration factor, F_{1-2} is then given by

$$F_{1-2} = \frac{1}{A_1} \int\limits_{A_1} \int\limits_{A_2} \frac{\cos(\theta_1) \cdot \cos(\theta_2)}{\pi S^2} dA_1 dA_2 \qquad (106)$$

61

When the surfaces A_1 and A_2 are far apart relative to their surface area, eq(106) can be approximated by assuming that θ_1, θ_2 and S are constant over the region of integration to obtain

$$F_{1-2} = \frac{\cos(\theta_1)\cos(\theta_2)}{\pi S^2} A_2 \tag{107}$$

The dot product form of the cosine defined in eqs (104) and (105) can be substituted into the previous equation to obtain

$$F_{1-2} = -\frac{(n_1 \cdot s)(n_2 \cdot s)}{\pi S^4} A_2 . \tag{108}$$

A simpler, though less accurate, approximation for the configuration factor can be made using the following observation. Suppose that surface 1 is a differential element at the center of a base of a hemisphere with area A_H and surface 2 is a region on this hemisphere with area A_2, then $F_{1-2} = A_2/A_H$. Therefore, if surface 1 is a differential element (*i.e.* our target) in a compartment and surface 2 is a wall in this compartment, then F_{1-2} can be approximated by

$$F_{1-2} = A_2/A_{total} \tag{109}$$

where A_2 is the area of the wall and A_{total} is the total area of the surfaces 'seen' by the target. The above equation rather than eq(108) is used to approximate the configuration factors.

Transmissivity: The transmissivity of a gas volume is the fraction of radiant energy that will pass through it unimpeded and is given by

$$\tau(y) = e^{-ay} \tag{110}$$

where a is the absorptance per unit length of the gas volume and y is a characteristic path length.

In a two layer zone model, a path between an object (fire, wall segment, *etc.*) and a target may traverse through both layers. In this case, the length of the path in the lower layer, y_L, can be computed given the total distance S between the object and target, and the elevations of the target, y_t, object, y_o and layer, y_{lay}, to be

$$y_L = \begin{cases} 0, & y_{lay} < y_{min} \\ \dfrac{y_{lay} - y_{min}}{y_{max} - y_{min}} S, & y_{min} \leq y_{lay} \leq y_{max} \\ S, & y_{lay} \geq y_{max} \end{cases} \qquad (111)$$

where

$$\begin{aligned} y_{min} &= \min(y_0, y_t) \\ y_{max} &= \max(y_0, y_t) \end{aligned} \qquad (112)$$

the path length in the upper layer is $y_U = S - y_L$, and the transmittance of the lower (upper) layer is denoted τ_L (τ_U).

Absorptivity: The absorptivity, α, of a gas volume is the fraction of radiant energy absorbed by that volume. For a grey gas $\alpha + \tau = 1$. The absorptivity of the lower (upper) layer is denoted α_L (α_U).

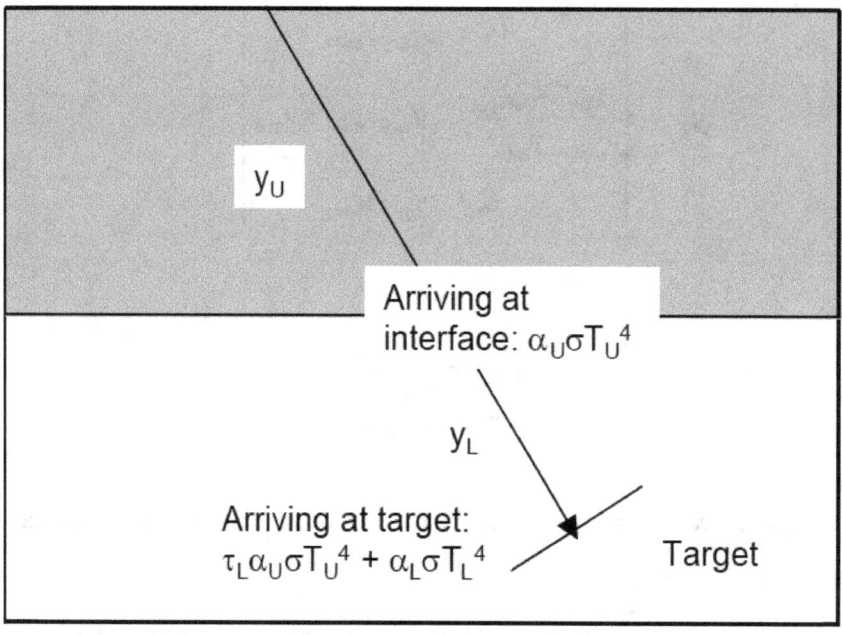

$\alpha_U = 1 - \exp(-a_U y_U) = $ emittance of upper layer

$\alpha_L = 1 - \exp(-a_L y_L) = $ emittance of lower layer

$\tau_L = \exp(-a_L y_L) = $ transmittance of lower layer

Figure 17. Radiative heat transfer from a wall surface in the upper layer to a target in the lower layer.

3.5.5.1 Computing the Heat Flux to a Target

There are four components of heat flux to a target: fires, walls, gas layer radiation and gas layer convection.

Heat Flux from a Fire to a Target: Figure 18 illustrates terms used to compute heat flux from a fire to a target. Let n_t be a unit vector perpendicular to the target and θ_t be the angle between the vectors $-s$ and n_t.

Using the definition that q_f is the radiative portion of the energy release rate of the fire, then the heat flux on a sphere of radius S due to this fire is $q_f /(4\pi S^2)$. Correcting for the orientation of the target and accounting for heat transfer through the gas layers, the heat flux to the target is

$$q_{f,t}'' = \frac{q_f}{4\pi S^2} \cos(\theta_t) \tau_U(y_u) \tau_L(y_L) = -q_f \frac{n_t \cdot S}{4\pi S^3} \tau_U(y_u) \tau_L(y_L) \tag{113}$$

Radiative Heat Flux from a Wall Segment to a Target: Figure 17 illustrates terms used to compute heat flux from a wall segment to a target. The flux, $q''_{w,t}$, from a wall segment to a target can then be computed using

$$q''_{w,t} = \frac{A_w q''_w(out) F_{w-t}}{A_t} \tau_U(y_U)\, \tau_L(y_L) \tag{114}$$

File Contains Data for
PostScript Printers Only

Figure 18. Radiative heat transfer from a point source fire to a target.

where $q''_w(out)$ is the flux leaving the wall segment, A_w, A_t are the areas of the wall segment and target respectively, F_{w-t} is the fraction of radiant energy given off by the wall segment that is intercepted by the target (*i.e.* a configuration factor) and $\tau_U(y_U)$ and $\tau_L(y_L)$ are defined as before. Equation (113) can be simplified using the symmetry relation $A_w F_{w-t} = A_t F_{t-w}$ (see [54], eq(7-25) or [55]) to obtain

$$q''_{w,t} = q''_w(out) F_{t-w} \tau_U(y_U)\tau_L(y_L) \tag{115}$$

where

$$q''_w(out) = \sigma T_w^4 - (1-\varepsilon_w)\frac{\Delta q''_w}{\varepsilon_w} \tag{116}$$

65

(see [54, eq (17-15)]) and $\sigma = 5.67^{-8}\ W/(m^2 K^4)$, T_w is the temperature of the wall segment, ϵ_w is the emissivity of the wall segment and $\Delta q_w''$ is the net flux striking the wall segment.

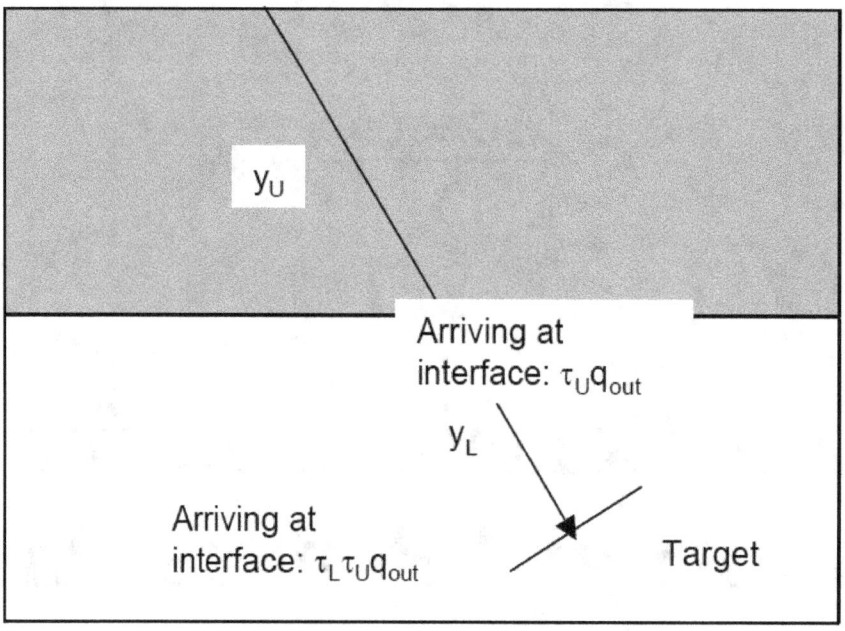

$\tau_U = \exp(-a_U y_U)$ = transmittance of upper layer

$\tau_L = \exp(-a_L y_L)$ = transmittance of lower layer

Figure 19. Radiative heat transfer from the upper and lower layer gas layers to a target in the lower layer.

Radiation from the Gas Layer to the Target: Figure 19 illustrates the setup for calculating the heat flux from the gas layers to the target. The upper and lower gas layers in a room contribute to the heat flux striking the target if the layer absorptances is non-zero. Heat transfer does not occur to the target when there are no fires and conditions are isothermal, i.e., target, wall segments and gas layers are at the same temperature. Therefore, substituting $q''_{f,t}=0$ into eq (98) we must have

$$\Delta'' q_t = 0 = \varepsilon_t \left(\sum_w q''_{w,t} + \sum_t q''_{q,t} - \sigma T_t^4 \right) + q''_{source} \qquad (117)$$

whenever $T_t = T_g = T_w$ for each wall segment and each gas layer.

66

Let $q''_{w,t}(gas)$ denote the flux striking the target due to the gas g in the direction of wall segment w. Then

$$q''_{w,t}(gas) = \begin{cases} \sigma F_{t-w}(T_L^4 \alpha_L \tau_U + T_U^4 \alpha_U) & \text{w is in the lower layer} \\ \sigma F_{t-w}(T_U^4 \alpha_U \tau_L + T_L^4 \alpha_L) & \text{w is in the upper layer} \end{cases} \tag{118}$$

The total target flux due to the gas (upper or lower layer) is obtained by summing eq (118) over each wall segment or

$$q''_{g,t} = \sum_w q''_{w,t}(gas). \tag{119}$$

The procedure expressed in eqs (118) and (119) for computing heat flux to the target from the gas layers satisfies the iso-thermal condition, eq (117), that no heat transfer occurs when all material have the same temperature. To see this, assume that $T_t = T_w = T_g$ and substitute eqs (115), (118) and (119) into eq (117) . Then $q''_{convec}=0$ and

$$\begin{aligned}
\sum_w q''_{w,t} + \sum_t q''_{g,t} &= \sum_{w\,upper} F_{t-w}\left(q''_w(out)\tau_U \tau_L + \sigma\left(T_t^4 \alpha_U \tau_L + T_t^4 \alpha_L\right)\right) + \\
&\quad \sum_{w\,lower} F_{t-w}\left(q''_w(out)\tau_L \tau_U + \sigma\left(T_t^4 \alpha_L \tau_U + T_t^4 \alpha_U\right)\right) \\
&= \sigma T_t^4 \sum_{w\,upper} F_{t-w}\left(\tau_U \tau_L + \alpha_U \tau_L + \alpha_L\right) + \\
&\quad \sigma T_t^4 \sum_{w\,lower} F_{t-w}\left(\tau_L \tau_U + \alpha_L \tau_U + \alpha_U\right) \\
&= \sigma T_t^4 \sum_w F_{t-w} = \sigma t_t^4
\end{aligned} \tag{120}$$

since $\tau_L \tau_U + \alpha_L \tau_U + \alpha_U = 1 = \tau_U \tau_L + \alpha_U \tau_U + \alpha_L$, $q_w''(out) = \sigma T_t^4$ ($\Delta q''_w$ in eq (119) is zero when conditions are iso-thermal) and $\Sigma_w F_{t-w} = 1$.

Convective Heat Transfer from a Gas Layer to a Target: The convective heat transfer from an adjoining gas layer to the target can be computed using existing CFAST routines. These routines require the orientation of the target, i.e., whether it is horizontal or vertical.

3.5.5.2 Computing Target Temperature

The steady state target temperature, T_t can be found by solving the equation $f(T_t) = 0$ where $f(T_t)$ is defined in eq (103). This can be done by using the Newton iteration

$$T_{new} = T_{old} - \frac{f(T_{old})}{f'(T_{old})} \tag{121}$$

where

$$f(T_t) = \epsilon_t \sigma T_t^4 - c(T_g - T_t) - \epsilon_t q''_{rad}(in)$$
$$f'(T_t) = \frac{d}{dT_t}\left(\epsilon_t \sigma T_t^4 - q''_{convec}\right) = 4\epsilon_t \sigma T_t^3 + c - \frac{dc}{dT_t}(T_g - T_t) \tag{122}$$

Note that $q''_{rad}(in)$ does not depend on the target temperature T_t so that $d/dT_t(q''_{rad}(in)) = 0$. If the convective heat transfer coefficient, c, in the above equation is independent of T_t then $d/dT_t(q''_{convec}) = -c$, otherwise (e.g. in the case of CFAST) this derivative may be evaluated numerically using finite differences. Equation (121) is iterated until the difference $T_{new} - T_{old}$ is sufficiently small.

3.5.6 Ceiling Jet

Relatively early in the development of a fire, fire-driven ceiling jets and gas-to-ceiling convective heat transfer can play a significant role in room-to-room smoke spread and in the response of near-ceiling mounted detection hardware. Cooper [56] details a model and computer algorithm to predict the instantaneous rate of convective heat transfer from fire plume gases to the overhead ceiling surface in a room of fire origin. The room is assumed to be a rectangular parallelopiped and, at times of interest, ceiling temperatures are simulated as being uniform. Also presented is an estimate of the convective heat transfer due to ceiling-jet driven wall flows. The effect on the heat transfer of the location of the fire within the room is taken into account. This algorithm has been incorporated into the CFAST model. In this section, we provide an overview of the model. Complete details are available in reference [56].

A schematic of a fire, fire plume, and ceiling jet is shown in Figure 20 The buoyant fire plume rises from the height Z_{fire} toward the ceiling. When the fire is below the layer interface, its mass and enthalpy flow are assumed to be deposited into the upper layer at height Z_{layer}. Having penetrated the interface, a portion of the plume typically continues to rise toward the ceiling. As it impinges on the ceiling surface, the plume gases turn and form a relatively high temperature, high velocity, turbulent ceiling jet which

68

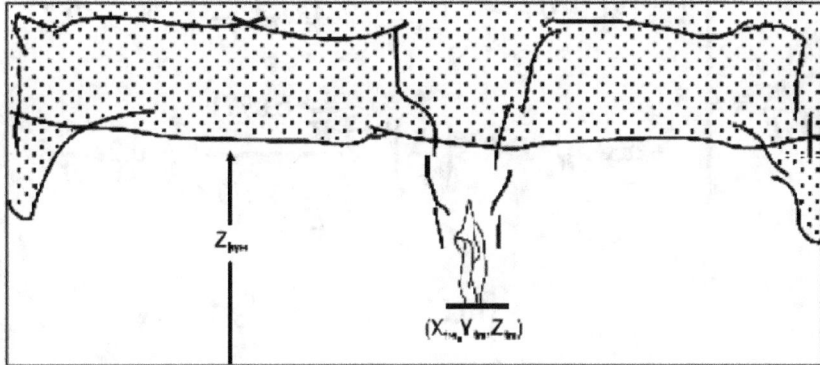

Figure 20. Convective heat transfer to ceiling and wall
surfaces via the ceiling jet.

flows radially outward along the ceiling and transfers heat to the relatively cool ceiling surface. The convective heat transfer rate is a strong function of the radial distance from the point of impingement, reducing rapidly with increasing radius. Eventually, the relatively high temperature ceiling jet is blocked by the relatively cool wall surfaces [57]. The ceiling jet then turns downward and outward in a compli cated flow along the vertical wall surfaces [58], [59]. The descent of the wall flows and the heat transfer from them are eventually stopped by upward buoyant forces. They are then buoyed back upward and mix with the upper layer.

The average convective heat flux from the ceiling jet gases to the ceiling surface, Q_{ceil}, can be expressed in integral form as

$$\dot{Q}_{ceil} = \int_0^{X_{wall}} \int_0^{Y_{wall}} \dot{q}''_{ceil}(x,y) \; dxdy \tag{123}$$

The instantaneous convective heat flux, $\dot{q}''_{ceil}(X,Y)$ can be determined as derived by Cooper [56]:

$$\dot{q}''_{ceil}(x,y) = h_1 \left(T_{ad} - T_{ceil} \right) \tag{124}$$

where T_{ad}, a characteristic ceiling jet temperature, is the temperature that would be measured adjacent to an adiabatic lower ceiling surface, and h_l is a heat transfer coefficient. h_l and T_{ad} are given by

$$\frac{h_l}{\hat{h}} = \begin{cases} 8.82 Re_H^{-1/2} Pr^{-2/3} \left(1 - \left(5 - 0.284 Re_H^{2/5}\right)\frac{r}{H}\right) & 0 \le \frac{r}{H} < 0.2 \\[3mm] 0.283 Re_H^{0.3} Pr^{-2/3} \left(\frac{r}{H}\right)^{-1.2} \dfrac{\frac{r}{H} - 0.0771}{\frac{r}{H} + 0.279} & 0.2 \le \frac{r}{H} \end{cases} \tag{125}$$

$$\frac{T_{ad} - T_u}{T_u Q_H^{*2/3}} = \begin{cases} 10.22 - 14.9 \frac{r}{H} & 0 \le \frac{r}{H} < 0.2 \\[3mm] 8.39 f\left(\frac{r}{H}\right) & 0.2 \le \frac{r}{H} \end{cases} \tag{126}$$

where

$$f\left(\frac{r}{H}\right) = \frac{1 - 1.10 \left(\frac{r}{H}\right)^{0.8} + 0.808 \left(\frac{r}{H}\right)^{1.6}}{1 - 1.10 \left(\frac{r}{H}\right)^{0.8} + 2.20 \left(\frac{r}{H}\right)^{1.6} + 0.690 \left(\frac{r}{H}\right)^{2.4}} \tag{127}$$

$$r = \left(\left(X - X_{fire}\right)^2 + \left(Y - Y_{fire}\right)^2\right)^{\frac{1}{2}} \tag{128}$$

$$\hat{h} = \rho_u c_p g^{1/2} H^{1/2} Q_H^{*1/3}; \qquad Re_H = \frac{g^{1/2} H^{3/2} Q_H^{*1/3}}{\nu_u}; \qquad Q_H^* = \frac{\dot{Q}'}{\rho_u c_p T_u (gH)^{1/2} H^2} \tag{129}$$

$$\dot{Q}' = \begin{cases} \dot{Q}_{fe} \dfrac{\sigma \dot{M}^*}{1 + \sigma} & Z_{fire} < Z_{layer} < Z_{ceil} \\[3mm] \dot{Q}_{fe} & \begin{array}{l} Z_{fire} \ge Z_{layer} \\ Z_{layer} = Z_{ceil} \end{array} \end{cases} \qquad \dot{M}^* = \begin{cases} 0 & -1 < \sigma \le 0 \\[3mm] \dfrac{1.04599\sigma + 0.360391\sigma^2}{1 + 1.37748\sigma + 0.360391\sigma^2} & \sigma > 0 \end{cases} \tag{130}$$

$$\sigma = \frac{1 - \frac{T_u}{T_l} + C_T Q_{EQ}^{*2/3}}{\frac{T_u}{T_l}}; \qquad C_T = 9.115 \tag{131}$$

70

$$Q_{EQ}^* = \left(\frac{0.21 Q_{fe}}{c_p T_l \dot{m}_p} \right)^{3/2}$$

(132)

In the above, H is the distance from the (presumed) point source fire and the ceiling, X_{fire} and Y_{fire} are the position of the fire in the room, Pr is the Prandtl number (taken to be 0.7) and v_u is the kinematic viscosity of the upper layer gas which is assumed to have the properties of air and can be estimated from $v_u = 0.04128(10^7) T_u^{5/2}/(T_u + 110.4)$. Q_H^* and Q_{EQ}^* are dimensionless numbers and are measures of the strength of the plume at the ceiling and the layer interface, respectively.

When the ceiling jet is blocked by the wall surfaces, the rate of heat transfer to the surface increases. Reference [56] provides details of the calculation of wall surface area and convective heat flux for the wall surfaces.

3.6 Detection

Detection is modeled using temperatures obtained from the ceiling jet[60]. Rooms without fires do not have ceiling jets. Sensors in these types of rooms use gas layer temperatures instead of ceiling jet temperatures. The characteristic smoke detector temperature is simply the temperature of the ceiling jet (at the location of the smoke detector). The characteristic heat detector temperature is modeled using the differential equation[61]

$$\frac{dT_L}{dt} = \frac{\sqrt{S(t)}}{RTI} \left(T_g(t) - T_L(t) \right)$$

(133)

$$T_L(0) = T_g(0)$$

(134)

where T_L, T_g are the link and gas temperatures, S is the flow speed of the gas and RTI (response time index) is a measure of the sensor's sensitivity to temperature change (thermal inertia). The heat detector differential eq (133) may be rewritten to

$$\frac{dT(t)}{dt} = a(t) - b(t) T(t)$$

(135)

$$T(t_0) = T_0$$

71

where

$$a(t) = \frac{\sqrt{S(t)}\, T(t)}{RTI}, \qquad b(t) = \frac{\sqrt{S(t)}}{RTI} \tag{137}$$

Equation (135) may be solved using the trapezoidal rule to obtain

$$\frac{T_{i+1} - T_i}{\Delta t} = \frac{1}{2}\left((a_i - b_i T_i) + (a_{i+1} - b_{i+1} T_{i+1}) \right) \tag{138}$$

where the subscript i denotes time at t_i. Equation (138) may be simplified to

$$T_{i+1} = A_{i+1} - b_{i+1} T_{i+1} \tag{139}$$

$$A_{i+1} = T_i + \frac{\Delta t}{2}\left(a_i - b_i T_i + a_{i+1} \right) \tag{140}$$

$$B_{i+1} = \frac{\Delta t}{2} b_{i+1} \tag{141}$$

which has a solution

$$T_{i+1} = \frac{A_{i+1}}{1 + B_{i+1}} = \frac{1 - \frac{\Delta t}{2} b_i}{1 + \frac{\Delta t}{2} b_{i+1}} T_i + \frac{\Delta t}{1 + \frac{\Delta t}{2} b_{i+1}}\left(\frac{a_i + a_{i+1}}{2} \right) \tag{142}$$

Equation (142) reduces to the trapezoidal rule for integration when $b(t) = 0$. When $a(t)$ and $b(t)$ are constant (the gas temperature, T_g, and gas velocity, S are not changing), eq (133) has the solution

72

$$T(t) = \frac{a}{b} + \frac{e^{-b(t-t_0)}(bT_0 - a)}{b} = T_g + e^{-\sqrt{A(t)}(t-t_0)/RTI}\left(T_0 - T_g\right) \qquad (143)$$

3.7 Suppression

For suppression, the sprinkler is modeled using a simple model [62] generalized for varying sprinkler spray densities [63]. Fire suppression in CFAST is then modeled by attenuating all fires in the room where the sensor activated by a term of the form $e^{-(t-tact)/trate}$ where t_{act} is the time when the sensor activated and t_{rate} is a constant determining how quickly the fire attenuates. The term t_{rate} can be related to spray density of a sprinkler using a correlation developed in [63]. The suppression correlation was developed by modifying the heat release rate of a fire. For $t>t_{act}$ the heat release is given by

$$\dot{q}_{fire}(t) = e^{-(t-t_{act})/3.0\, \dot{q}_{spray}^{-1.8}}\, \dot{q}(t_{act})$$

where q_{spray} is the spray density of a sprinkler. Note that decay rate can be formulated in terms of either the attenuation rate or the spray density. Both options are available. t_{rate} can be expressed in terms of q_{spray} as

$$t_{rate} = 3.0\, \dot{q}_{spray}^{-1.8} \qquad (145)$$

$$t_{50\%} = 3\ln(2)\, \dot{q}_{spray}^{-1.8} \qquad (146)$$

and the decay time (time to 50% attenuation) as the input line allows the specification of either the spray density of the sprinkler q_{spray} or the time required to reduce the fire release rate by 50%, $t_{50\%}$. The chemistry routine performs the combustion chemistry, making sure that the fuel burned is consistent with the available oxygen. If detection has occurred then the mass and energy release rates are attenuated by the term $e^{-(t-tact)}=t_{rate}$ to obtain

$$\dot{m}_{pyrols}(t) = e^{-(t-t_{act})/t_{rate}}\, \dot{m}_{pyrols}(t_{act}) \qquad (147)$$

73

$$m_{pyrols}(t) = e^{-(t-t_{act})/t_{rate}} m_{pyrols}(t_{act}). \tag{148}$$

Another approach would be to calculated the $m_{pyrols}(t)$ mass production rate that would result in the desired constrained energy release rate computed in CHEMIE. This is not practical since the energy release rate $q_{chemie}(t)$ is a non-linear function (via the plume entrainment function) of the pyrolis rate $m_{pyrols}(t)$.

There are assumptions and limitations in this approach. Its main deficiency is that it assumes that sufficient water is applied to the fire to cause a decrease in the rate of heat release. This suppression model cannot handle the case when the fire overwhelms the sprinkler. The suppression model as implemented does not include the effect of a second sprinkler. Detection of all sprinklers are noted but their activation does not make the fire go out any faster. Further, multiple fires in a room imply multiple ceiling jets. It is not clear how this should be handled, ie how two ceiling jets should interact. When there is more than one fire, the detection algorithm uses the fire that results in the worst conditions (usually the closest fire) in order to calculate the fire sensor temperatures. The ceiling jet algorithm that we use results in temperature predictions that are too warm (as compared to ceiling jet full scale experiments of Madrzykowski). This has not been resolved.

3.8 Species Concentration and Deposition

3.8.1 Species Transport

The species transport in CFAST is really a matter of bookkeeping to track individual species mass as it is generated by a fire, transported through vents, or mixed between layers in a compartment. When the layers are initialized at the start of the simulation, they are set to ambient conditions. These are the initial temperature specified by the user, and 23 % by mass (21 % by volume) oxygen, 77 % by mass (79 % by volume) nitrogen, a mass concentration of water specified by the user as a relative humidity, and a zero concentration of all other species. As fuel is burned, the various species are produced in direct relation to the mass of fuel burned (this relation is the species yield specified by the user for the fuel burning). Since oxygen is consumed rather than produced by the burning, the "yield" of oxygen is negative, and is set internally to correspond to the amount of oxygen used to burn the fuel (within the constraint of available oxygen limits discussed in sec. 3.1.2).

Each unit mass of a species produced is carried in the flow to the various rooms and accumulates in the layers. The model keeps track of the mass of each species in each layer, and knows the volume of each layer as a function of time. The mass divided by the volume is the mass concentration, which along with the molecular weight gives the concentration in volume % or ppm as appropriate.

For soot, the input for C/CO$_2$ is used to calculate a "soot" yield from the fire (assuming all the carbon goes to soot). This soot generation is then transported as a species to yield a soot mass concentration to use in the optical density calculation based on the work of Seader and Einhorn[64].

3.8.2 HCl Deposition

Hydrogen chloride produced in a fire can produce a strong irritant reaction that can impair escape from the fire. It has been shown [65] that significant amounts of the substance can be removed by adsorption by surfaces which contact smoke. In our model, HCl production is treated in a manner similar to other species. However, an additional term is required to allow for deposition on, and subsequent absorption into, material surfaces.

The physical configuration that we are modeling is a gas layer adjacent to a surface (Figure 21). The gas layer is at some temperature T_g with a concomitant density of hydrogen chloride, ρ_{HCl}. The mass transport coefficient is calculated based on the Reynolds analogy with mass and heat transfer: that is, hydrogen chloride is mass being moved convectively in the boundary layer, and some of it simply sticks to the wall surface rather than completing the journey during the convective roll-up associated with eddy diffusion in the boundary layer. The boundary layer at the wall is then in equilibrium with the wall. The latter is a statistical process and is determined by evaporation from the wall and stickiness of the wall for HCl molecules. This latter is greatly influenced by the concentration of water in the gas, in the boundary layer and on the wall itself.

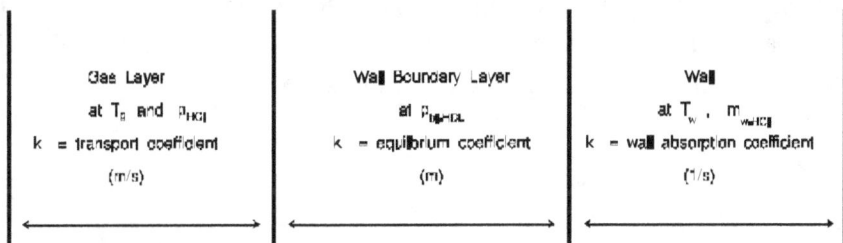

Figure 21. Schematic of hydrogen chloride deposition region.

The rate of addition of mass of hydrogen chloride to the gas layer is given by

$$\frac{d}{dt}m_{HCl} = source - k_\varepsilon \times \left(\rho_{HCl} - \rho_{bHCl}\right) \times A_w \qquad (149)$$

where source is the production rate from the burning object plus flow from other compartments.

75

For the wall concentration, the rate of addition is

$$\frac{d}{dt} d_{HCl,w} = k_e \times \left(\rho_{HCl} - \rho_{blHCl} \right) - k_s \times m_{HCl,w} \tag{150}$$

where the concentration in the boundary layer, ρ_{blHCl}, is related to the wall surface concentration by the equilibrium constant k_e,

$$\rho_{blHCl} = d_{HCl,w} / k_e . \tag{151}$$

We never actually solve for the concentration in the boundary layer, but it is available, as is a boundary layer temperature if it were of interest. The transfer coefficients are

$$k_e = \frac{\dot{q}}{\Delta T \, \rho_g \, c_p} \tag{152}$$

$$k_s = \frac{b_1 \, e^{1500/T_w}}{1 + b_2 \, e^{1500/T_w} \, \rho_{hcl}} \left(1 + \frac{b_5 \left(\rho_{H_2O} \right)^{b6}}{\left(\rho_{H_2O,sat} - \rho_{H_2O,g} \right)^{b7}} \right) \tag{153}$$

$$k_s = b_3 \, e^{-\left(\frac{b_4}{R \, T_w} \right)} . \tag{154}$$

The only values currently available [66] for these quantities are shown in table 4. The "b" coefficients are parameters which are found by fitting experimental data to eqs (149) through (154). These coefficients reproduce the adsorption and absorption of HCl reasonably well. Note though that error bars for these coefficients have not been reported in the literature.

Table 4. Transfer coefficients for HCl deposition

Surface	b_1 (m)	b_2 (m³/kg)	b_3 (s⁻¹)	b_4 (J/g-mol)	b_5 (note a)	b_6 (note b)	b_7 (note c)
Painted Gypsum	0.0063	191.8	0.0587	7476.	193	1.021	0.431
PMMA	9.6×10^{-5}	0.0137	0.0205	7476.	29	1.0	0.431
Ceiling Tile	4.0×10^{-3}	0.0548	0.123	7476.	30[b]	1.0	0.431
Cement Block	1.8×10^{-2}	5.48	0.497	7476.	30[b]	1.0	0.431
Marinite®	1.9×10^{-2}	0.137	0.030	7476.	30[b]	1.0	0.431

a units of b_5 are $(m^3/kg)^{\wedge}(b_7 - b_6)$
b very approximate value, insufficient data for high confidence value
c non-dimensional

The experimental basis for poly(methyl methacrylate) and gypsum cover a sufficiently wide range of conditions that they should be usable in a variety of practical situations. The parameters for the other surfaces do not have much experimental backing, and so their use should be limited to comparison purposes.

3.9 Flame Spread

The Quintiere-Cleary flame spread model incorporated into CFAST is based on five simple differential equations. One each for concurrent, eq (155), and opposed flow flame spread, eq (156). One each, eqs (158) and (157) for the two burnout fronts and the last one for burn out at the ignition point eq (159).

$$\frac{dy_p}{dt} = \frac{y_f - y_p}{\frac{1}{4}k\rho c[\frac{T_{ig}-T_s}{\dot{q}_f''}]^2} \tag{155}$$

$$\frac{dx_p}{dt} = \frac{\phi}{k\rho c(T_{ig}-T_s)^2}, \text{ for } T_s \geq T_{s,min} \tag{156}$$

77

$$\frac{dy_b}{dt} = \frac{Q''(y_p - y_b)}{Q''_{TOT}} \tag{157}$$

$$\frac{dx_b}{dt} = \frac{Q''(x_p - x_b)}{Q''_{TOT}} \tag{158}$$

and

$$\frac{dQ''}{dt} = \frac{(q''_f - \sigma T_{ig}^4 + \sigma T_{Layer}^4)}{\Delta L} \Delta H \tag{159}$$

The equations describe the growth of two rectangles. At ignition a single rectangle, R_p, is defined and its growth is determined by eq (155) for spread up the wall and eq (156) for lateral spread as well as spread down the wall. When $Q'' \leq 0$ a second rectangle, R_b, the same size as R_p was originally starts growing. It is governed by eqs (158) and (157). After R_b starts to be tracked the pyrolysis area is $R_p - R_b$.

When a flame spread object is defined, CFAST adds five additional differential equations to the equation set. A target is also placed at the specified location on the specified wall surface and the maximum time step is set to be 1 s. This allows the temperature of the target to be tracked and the ignition temperature and time to be calculated,

Once a flame spread object ignites, its mass loss and heat release rate are calculated and then treated like any other object fire by CFAST.

3.10 Single Zone Approximation

A single zone approximation may be derived by using two-zone source terms and the substitutions:

78

$$\dot{m}_U^{new} = \dot{m}_L + \dot{m}_U,$$
$$\dot{m}_L^{new} = 0$$
$$\dot{q}_U^{new} = \dot{q}_L + \dot{q}_U,$$
$$\dot{q}_L^{new} = 0. \tag{160}$$

This is not a fundamental improvement, but rather is designed to fit in with the concept of single zone and network models that are being utilized currently.

3.11 Miscellaneous Phenomena

Several improvements have been incorporated into the fire model based on experience in using it. One was to include the calculation of the species oxygen in the basic equation set. As is discussed elsewhere, the basic equations are for the upper and lower layer temperatures, the upper layer volume and the pressure at the floor of the compartment. These equations are derived from fluid dynamics and are based on the conservation of mass, momentum and energy. They can form a set of stiff, ordinary differential equations. In general, however, it has been a common practice to assume that the phenomena that drive these flows happen either much more quickly or much more slowly than that characteristic of the fluid flow time scales. In the case of fires where lack of oxygen can limit the combustion, and therefore the driving force, the oxygen which flows into a fire is intimately coupled to the forcing of the fluid flow and therefor the set becomes stiff.

4 Structure of the Model

In this chapter, details of the implementation of the model are presented. These include

- an overview of the model formulation,
- the structure of the model including the major routines implementing the various physical phenomena included in the model,
- the organization of data initialization and data input used by the model,
- the structure of data used to formulate the differential equations solved by the model,
- a summary of the main control routines in the model that are used to control all input and output, initialize the model and solve the appropriate differential equation set for the problem to be solved, and
- guidelines for modifying the model to include new or enhanced physical phenomena.

4.1 Subroutine Structure

The model can be split into distinct parts. There are routines for reading data, calculating results and reporting the results to a file or printer. The major routines for performing these functions are identified in Figure 22. These physical interface routines link the CFAST model to the actual routines which calculate quantities such as mass or energy flow at one particular point in time for a given environment.

The routines SOLVE, RESID and DASSL are the key to understanding how the physical equations are solved. SOLVE is the control program that oversees the general solution of the problem. It invokes the differential equation solver DASSL [67] which in turn calls RESID to solve the transport equations. The problem that these routines solve is as follows. Given a solution at time t, what is the solution at time $t + \Delta t$? The differential equations are of the form

$$\frac{dy}{dt} = f(y,t)$$
$$y(t_0) = y_0$$

(161)

where y is a vector representing pressure, layer height, mass, *etc.* and f is a vector function that represents changes in these values with respect to time. The term y_0 is an initial condition at the initial time t_0. The subroutine RESID (see sec. 4.3) computes the right hand side of eq (161) and returns a set of residuals of that calculation to be compared to the values expected by DASSL. DASSL then checks for convergence. Once DASSL reaches an error limit (defined as convergence of the equations) for the solution at $t+\Delta t$, SOLVE then advances the solution of species concentration, wall temperature profiles, and mechanical ventilation for the same time interval.

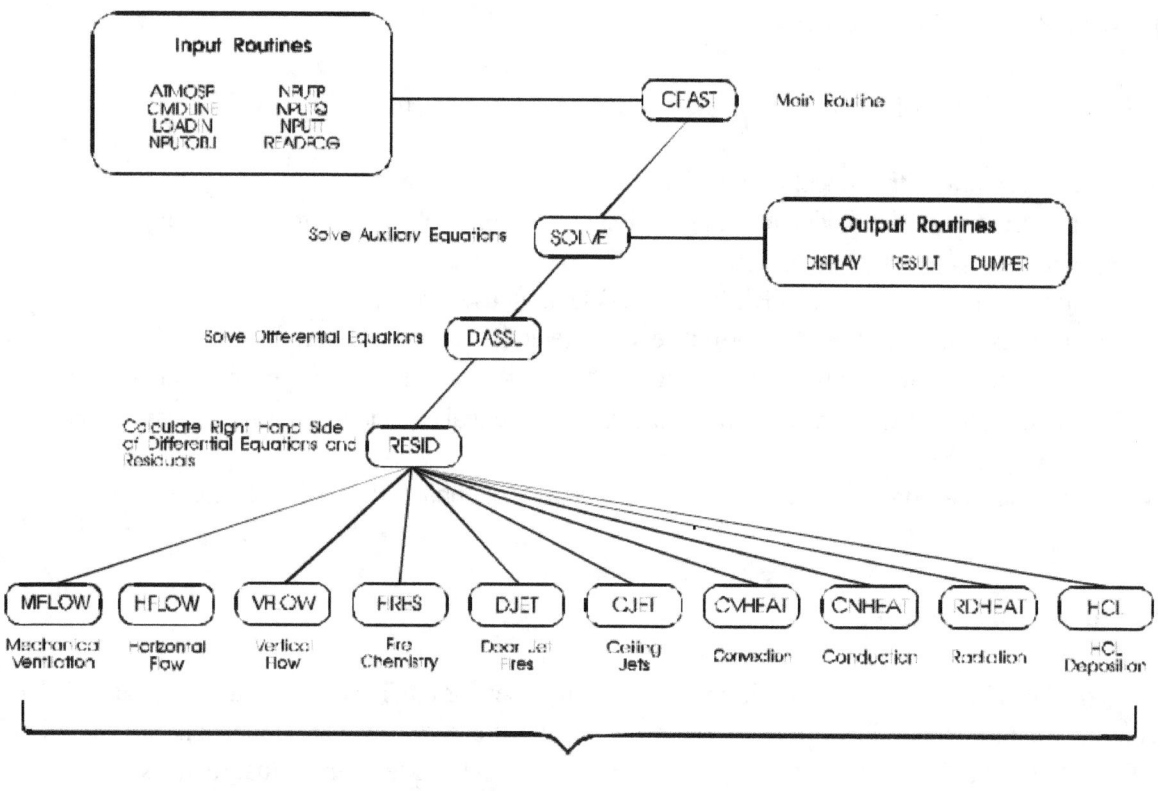

Figure 22. Subroutine structure for the CFAST Model.

Note that there are several distinct time scales that are involved in the solution of this type of problem. The fastest will be chemical kinetics. We avoid that scale by assuming that the chemistry is infinitely fast. The next larger time scale is that associated with the flow field. These are the equations which are cast into the form of ordinary differential equations. Then there is the time scale for mechanical ventilation, and finally, heat conduction through objects. By way of example, chemical kinetic times are typically on the order of milliseconds. The transport time scale will be on the order of 0.1 s. The mechanical ventilation and conduction time scales are typically several seconds, or even longer. Unlike earlier versions of the model, this new version dynamically adjusts the time step over the entire simulation to a value appropriate for the solution of the currently defined equation set. In addition to allowing a more correct solution to the pressure equation, very large time steps are possible if the problem being solved approaches steady-state.

4.2 Data Flow

4.2.1 Input and Initialization

Command Line and Program Options: Each main module in the CFAST suite calls a set of routines which set up the physical environment for the model. The routines are READOP and OPENSHEL. They perform a number of housekeeping tasks. This information is related to the environment (computer platform) on which the model is running.

First, the command line is interpreted, providing file names and options. Up to two file names are available. The first entry is either an input file, or a configuration file. If it is an input file, then the name is provided in the variable "NNFILE" in the shell common block. It will not be opened. When a file is opened for input, the unit number IOFILI should be assigned. If there is a valid output file, its name is provided in the variable "OUTFILE." It will be opened, and the unit number is IOFILO. There is a default configuration file HV1.CF which will be read unless an alternative is provided on the command line. A configuration file name can be provided in place of the input file. In this case, the input file will be fetched from the variable DFILE from within the configuration file. In any case, the current configuration file will be named in CONFIG. Most modules will not run without the configuration file.

Second, the options are stored in the shell common blocks. There are currently four options available. All five are available to all modules. They are specified on the command line by

> - (or /) option.

An example is the option to prevent the header from printing. This would be

> -N.

The four which are presently read and decoded are

> 1. Report type (Rnn)
> 2. No header (N)
> 3. Turn on error logging (L)
> 4. Pass an environment file (Ffilename) - *used internally by CFAST to pass an environment file between modules*

These options are read and interpreted by the routine READOP.

Input Data File Processing: The data file is opened by the input routine, NPUTP. It is closed by this routine after all data has been read. There should be no units assigned, opened or closed while data retrieval is in process. Thus data entry should be done within the scope of NPUTP or NPUTQ only by these routines. Any data can be retrieved from the data files by NPUTQ. If subsidiary information is needed, then a reference file name should be read and stored in a variable kept in the PARAMS common block (or unlabeled common), and data fetches made after the *second* exit from NPUTP. This precludes initialization of such data during the geometry and fire specification process. The thermal properties are retrieved by the initialization routines at this point (after the restart return).

New key words that are to be added to the data file are placed in NPUTQ. It is important to follow the protocol as laid out therein, so that consistency checks can be performed on the data. Any physical initialization that needs to be done should be included in NPUTP, after the call to NPUTG. At this point, the model is completely set up. The only data that is not done are the total masses of the upper and lower layers. This is deferred to the originating routine. The name of the primary data file is in the variable NNFILE, but is open at this point, and pointing to the end of NNFILE. NPUTP can be referenced twice if a restart has been requested. In each case, NNFILE is closed prior to exiting.

In the process of inserting key words into NPUTQ, one will note that there are two case statements for the key words. The first is for a normal start, and the second is for a restart. In the latter case, some variables are not, and should not, be valid. An example of the difference: the fire specification can change, but it makes no sense to change the physical layout, such as the number of compartments.

The next consideration is the setup provided for the physical system. Initialization is done by CFAST, or INITFS in the case of the data editor. Both preset memory, and call the routine NPUTP. NPUTP does the actual physical initialization. It in turn calls the routine NPUTQ which reads the data files. Subsequently, the geometry, species and graphics descriptors are set. Finally the environment is set by CFAST or INITFS. This includes reading the thermophysical properties and assigning them to the correct boundary, and all other auxiliary files as necessary. It is at this point that all the data files are closed, and the unit IOFILI is available. It should be closed, opened to the appropriate file, and *subsequently closed* if used for other purposes.

Problem Initialization: There are several logical switches that are set, based on the problem to be solved. The two most common are ACTIVS and SWITCH. The former is for active species. The latter serves two purposes, for active conduction and for miscellaneous parameters.

If a species is being computed, for whatever reason, then ACTIVS will be TRUE, otherwise it will be FALSE. This parameter is dimensioned to NS, the number of species which CFAST will follow. The order is shown in table 5.

For each species that is tracked, the variable ACTIVS(i) is set to true. There are two types of action that hinge on the setting of this variable. The first is in the availability and display of species information. The second is in the packing used in preparing the source terms for, and extracting them from the solver. The details of this activity are in the section on the data copy and RESID routines.

Table 5. Indices for species tracked by CFAST

INDEX	SPECIES	APPLICABLE KEY WORD
1	Nitrogen	none
2	Oxygen	O2
3	Carbon dioxide	none
4	Carbon monoxide	CO
5	Water vapor	none but HCR is related
6	Hydrogen chloride	HCL
7	Unburned hydrocarbons	none
8	Hydrogen cyanide	HCN
9	Soot	OD
10	Concentration time dose	CT (not a species)

The variable SWITCH is used in two places. The first is to specify which boundaries in which compartments can conduct heat and where HCl deposition takes place. The parameter is set in NPUTQ, but verified by NPUTT. It can be set in NPUTQ if specified in the data file, but subsequently turned off by NPUTT if the name of the boundary can not be found in the thermophysical database. When the primary model is running, it will terminate if this latter condition is found, whereas the data editor will distinguish between the boundary being considered adiabatic with the name "OFF" and not found by "NONE." SWITCH is dimensioned NWAL by NR.

Since conduction is only allowed for NR-1 compartments, the last column can be used for miscellaneous variables. Once again, the default is false, but if the appropriate key word has been set, then the variable will be set to true.

(1,NR) - print the flow field and species - set in NPUTP; used only by the main model

The order of initialization is important. This is particularly true because of the caveat above that input/output units should not be assigned during the primary initialization. First, the main routine, CFAST or INITFS, initialize memory, and some physical constants such as the gravitational constant. The main data file is then opened. If one can not be found, then the model quits. The next step is to call NPUTP with the restart parameter of ISRSTR=1. NPUTP reads the header line to check for a correct file, then calls NPUTQ, with the ISRSTR=1. NPUTQ does all of the actual data entry, *via* unit IOFILI. After control is returned to NPUTP, physical initialization is done, for example setting the atmospheric ambient, calculating the volume of the compartments, and so forth. Then the graphics descriptors are read by LOADIN. These processes occur whether or not a restart will be done. Then control is returned to the main module. Some additional processing takes place to set the species of the ambient environment. If a restart has been requested, the appropriate history file is read for the requested interval. NPUTP rewinds the input file and once again calls NPUTQ, with ISRSTR=2. At this point there are some differences. Within NPUTQ, the case statement (discussed above) prevents some parameters from being reset.

Units are specific to the operating system. There are two general input/output units named IOFILI and IOFILO. In the current implementation they are numbered 1 and 6 respectively. There are additional units as follows:

1) ISRSTR tells us if this is a normal input file, or a restart

        0 =>   Most data has been set - do initialization only
        1 =>   Implies a normal read, with open
        2 =>   Implies an update after restart

2) files used

        1 =>   Configuration file, primary data file and data bases - opened and closed
                   by each routine
        2 =>   Help files
        3 =>   Log file - open all the time
        9 =>   History file i/o & font files - both are open intermittently
      98 =>   ASCII history output
      99 =>   "Other Objects" database

4.2.2 Data Structure and Data Flow Within CFAST

Information is passed between the various subroutines and main modules by the use of files and common blocks. The model has several common blocks associated with it. Of interest to most programmers is the way these data are used in the model. Appendices A through D provide details of the subroutines and variables in the CFAST model.

4.3 The Control Programs (SOLVE and RESID)

As discussed above, the routine RESID controls most of the model calculations. SOLVE coordinates the solution and output, but the physical phenomena are accessed by RESID. This section provides an annotated overview of these two control programs.

4.3.1 <u>SOLVE</u>

Since much of the function of the control routine SOLVE is bookkeeping, the source code is not particularly illustrative. Rather, for this routine, we will provide an summary of the functioning of the routine below.

1. Initialize the print, history, and plot times to the user's input specifications.

2. By calling routine INITSOLN, determine a set of initial pressures consistent with the initial conditions (temperatures and vent sizes) of the problem to be solved. By solving a set of linear equations to determine appropriate steady-state initial conditions for the pressures, the differential equation solver is able to determine solutions for the always difficult first second of the solution several times faster than allowing the differential equation solver to find the initial solution.

3. Output results (or initial conditions at time t=0) of the calculation by printing (routine RESULT), writing a history file interval (routine DUMPER), or plotting (routine DISPLAYC) results if current time is appropriate for such output.

4. Call the differential equation solver, DASSL, to advance the solution in time. The length of the advance in time is chosen dynamically by DASSL. DASSL chooses the time step but reports back a solution based on the lesser of the print, display and dump intervals. DASSL call RESID to compute the actual solution, as well as the residuals.

5. Advance the solution for species not handled directly by the differential equation solver by calling routines RESID (again, but with different switches) and TOXIC.

6. Repeat steps 3-5 until the final time is reached.

Note that many of the defaults, switches, and tolerances can be changed in the initialization routine. SOLVE calls the initialization routine INITSLV. It has switches built in which can be redefined with the configuration file, SOLVER.INI. If this file is not present, then internal defaults are used which are the best available. However, for testing, turning off phenomena, alternative settings can be useful. See the appendices at the end for the format of this configuration file.

4.3.2 <u>RESID</u>

RESID is split into several parts. First, the current environment is copied from the form used by the differential equation solver into the environment common blocks for use by the physical routines in CFAST. Then the physical phenomena are calculated with calls to appropriate physical interface routines. Each physical interface routine returns its contribution to mass, enthalpy, and species flows into each layer in each room. These are then summed into total mass, enthalpy, and species flows into each layer in each room. Finally, the differential equations are formed for each room, wall surface, and mechanical ventilation system in the problem.

This portion of the model is the real numerical implementation and is accessed many times per simulation run. Careful thought must be given to the form of the routines since the execution time is *very* sensitive to the coding of the software. What follows is an annotated form of the routine RESID. Extraneous comments have been left out to shorten the listing somewhat.

```
SUBROUTINE RESID(TSEC,X,XPSOLVE,DELTA,IRES,RPAR,IPAR)
```

Common blocks go here to define the environment for CFAST use by all physical routines. Definition of temporary variables used to store the output of each physical routine are also included here. See section 4.4 for the format of each of these variables. The routine DATACOPY is called to copy the environment from the form used by the differential equation solver into the environment common blocks for use by the physical routines in CFAST.

```
      XX0 = 0.0D0
      ND = 0
      NPROD = NLSPCT
      DT = TSEC - TOLD
C
      NIRM = NM1
C
      CALL DATACOPY(X,ODEVARA+ODEVARB)
```

The IPAR and RPAR parameters are passed from SOLVE to RESID via DASSL and are used to control the calculation of the residuals by RESID. For a call to RESID from the differential equation solver, DASSL, IPAR(2) is equal to the parameter SOME to indicate that the routine is to calculate the set of differential equations without including the species. Species are updated by SOLVE once DASSL has found an appropriate solution for the smaller equation set.

```
      IF (IPAR(2).EQ.SOME) THEN
        UPDATE = .FALSE.
      ELSE
        UPDATE = .TRUE.
      END IF
      EPSP = RPAR(1)
```

All of the physical phenomena included in the model are included here with calls to the physical interface routines for each phenomena. Each physical interface routine returns its contribution to mass, enthalpy, and species flows into each layer in each room.

```
C
C      CALCULATE FLOW THROUGH VENTS (HFLOW FOR HORIZONTAL FLOW
C      THROUGH VERTICAL VENTS, VFLOW FOR VERTICAL FLOW THROUGH
C      HORIZONTAL VENTS, AND MVENT FOR MECHANICAL VENTILATION)
C
       CALL HFLOW(TSEC,EPSP,NPROD,FLWNVNT,QLPQUV)
       CALL VFLOW(FLWHVNT,QLPQUH)
       CALL MVENT(X(NOFPMV+1),X(NOFTMV+1),XPSOLVE(NOFTMV+1),FLWMV,
      +      DELTA(NOFPMV+1),DELTA(NOFTMV+1),XPRIME(NOFHVPR+1),NPROD)
C
C      CALCULATE HEAT AND MASS FLOWS DUE TO FIRES
C
       CALL FIRES(TSEC,FLWF,QLPQUF,NFIRE,IFROOM,XFIRE)
       CALL SORTFR(NFIRE,IFROOM,XFIRE,IFRPNT)
       CALL DJET(NFIRE,FLWDJF,XFIRE)
C
C      CALCULATE FLOW AND FLUX DUE TO HEAT TRANSFER (CEILING JETS,
C      CONVECTION AND RADIATION
C
       CALL CJET(IFRPNT,XFIRE,ND,XD,YD,ZD,FLWCJET,FLXCJET,TD,VD)
       CALL CVHEAT(IFRPNT,FLWCV,FLXCV)
       CALL RDHEAT(IFRPNT,XFIRE,FLWRAD,FLXRAD)
C
C      CALCULATE HCL DEPOSITION TO WALLS
C
       CALL HCL(FLWHCL,FLXHCL)
```

The flows returned from each physical interface routine are then summed into total mass, enthalpy, and species flows into each layer in each room. The form of each of these flows is discussed in section 4.4. In general, the array FLWTOT(room,species,layer) contains the total flow of each species into each layer of each room in the simulation. For ease of definition, mass and enthalpy are included in this array as pseudo-species (1 & 2) and summed along with the actual species (3 to 2+lsp). Heat flux to surfaces is included in a similar manner for used by the conduction routine.

```
C
C      SUM FLOW FOR INSIDE ROOMS
C
       DO 50 IROOM = 1, NIRM
         QLPQUR(IROOM) = QLPQUV(IROOM) + QLPQUH(IROOM) + QLPQUF(IROOM) +
      +      FLWCV(IROOM,LL) + FLWCV(IROOM,UU) + FLWRAD(IROOM,LL) +
      +      FLWRAD(IROOM,UU) + FLWCJET(IROOM,LL) + FLWCJET(IROOM,UU) +
      +      FLWDJF(IROOM,Q,LL) + FLWDJF(IROOM,Q,UU) +
      +      FLWMV(IROOM,Q,LL) + FLWMV(IROOM,Q,UU)
         DO 40 IPROD = 1, NPROD + 2
           IP = IZPMAP(IPROD)
           FLWTOT(IROOM,IPROD,LL) = FLWNVNT(IROOM,IPROD,LL) +
      +        FLWMV(IROOM,IP,LL) + FLWF(IROOM,IP,LL) +
      +        FLWDJF(IROOM,IP,LL) + FLWHVNT(IROOM,IP,LL)
           FLWTOT(IROOM,IPROD,UU) = FLWNVNT(IROOM,IPROD,UU) +
      +        FLWMV(IROOM,IP,UU) + FLWF(IROOM,IP,UU) +
      +        FLWDJF(IROOM,IP,UU) + FLWHVNT(IROOM,IP,UU)
  40     CONTINUE
C
C      ADD IN HCL CONTRIBUTION TO FLWTOT
C
         IF (ACTIVS(6)) THEN
           FLWTOT(IROOM,1,LL) = FLWTOT(IROOM,1,LL) + FLWHCL(IROOM,1,LL)
           FLWTOT(IROOM,1,UU) = FLWTOT(IROOM,1,UU) + FLWHCL(IROOM,1,UU)
           FLWTOT(IROOM,8,LL) = FLWTOT(IROOM,8,LL) + FLWHCL(IROOM,8,LL)
           FLWTOT(IROOM,8,UU) = FLWTOT(IROOM,8,UU) + FLWHCL(IROOM,8,UU)
         END IF
```

```
C
          FLWTOT(IROOM,Q,LL) = FLWTOT(IROOM,Q,LL) + FLWCV(IROOM,LL) +
     +          FLWRAD(IROOM,LL) + FLWCJET(IROOM,LL)
          FLWTOT(IROOM,Q,UU) = FLWTOT(IROOM,Q,UU) + FLWCV(IROOM,UU) +
     +          FLWRAD(IROOM,UU) + FLWCJET(IROOM,UU)
C
  50  CONTINUE
C
C    SUM FLUX FOR INSIDE ROOMS
C
      DO 70 IROOM = 1, NIRM
        DO 60 IWALL = 1, NWAL
          IF (SWITCH(IWALL,IROOM)) THEN
            FLXTOT(IROOM,IWALL) = FLXCV(IROOM,IWALL) +
     +            FLXRAD(IROOM,IWALL) + FLXCJET(IROOM,IWALL)
          END IF
  60    CONTINUE
  70  CONTINUE
```

The differential equations are formed for each room, wall surface, and mechanical ventilation system in the problem. These follow directly from the derivation in section 2.2.

```
      DO 80 IROOM = 1, NIRM
        AROOM = AR(IROOM)
        HCEIL = HR(IROOM)
        PABS = ZZPABS(IROOM)
        HINTER = ZZHLAY(IROOM,LL)
        QL = FLWTOT(IROOM,Q,LL)
        QU = FLWTOT(IROOM,Q,UU)
        TMU = FLWTOT(IROOM,M,UU)
        TML = FLWTOT(IROOM,M,LL)
        QLPQU = QLPQUR(IROOM)
C
C    PRESSURE EQUATION
C
        PDOT = (GAMMA-1.0D0) * QLPQU / (AROOM*HCEIL)
        XPRIME(IROOM) = PDOT
C
C    UPPER LAYER TEMPERATURE EQUATION
C
        TLAYDU = (QU-CP*TMU*ZZTEMP(IROOM,UU)) / (CP*ZZMASS(IROOM,UU))
        IF (OPTION(FODE).EQ.ON) THEN
          TLAYDU = TLAYDU + PDOT / (CP*ZZRHO(IROOM,UU))
        END IF
        XPRIME(IROOM+NOFTU) = TLAYDU
C
C    UPPER LAYER VOLUME EQUATION
C
        VLAYD = (GAMMA-1.0D0) * QU / (GAMMA*PABS)
        IF (OPTION(FODE).EQ.ON) THEN
          VLAYD = VLAYD - ZZVOL(IROOM,UU) * PDOT / (GAMMA*PABS)
        END IF
        XPRIME(IROOM+NOFVU) = VLAYD
C
C    LOWER LAYER TEMPERATURE EQUATION
C
        TLAYDL = (QL-CP*TML*ZZTEMP(IROOM,LL)) / (CP*ZZMASS(IROOM,LL))
        IF (OPTION(FODE).EQ.ON) THEN
          TLAYDL = TLAYDL + PDOT / (CP*ZZRHO(IROOM,LL))
        END IF
        XPRIME(IROOM+NOFTL) = TLAYDL
  80  CONTINUE
```

The species are only calculated once DASSL has an acceptable solution for the equation set not including the species. We presume that the species production rates occur on a time scale similar to the total mass production which is solved directly.

```
      IF (NPROD.GT.0.AND.IPAR(2).EQ.ALL) THEN
        IPRODU = NOFPRD - 1
        DO 100 IPROD = 1, NPROD
          DO 90 IROOM = 1, NM1
            HCEIL = HR(IROOM)
            HINTER = ZZHLAY(IROOM,LL)
            IPRODU = IPRODU + 2
            IPRODL = IPRODU + 1
            IF (HINTER.LT.HCEIL) THEN
              XPRIME(IPRODU) = FLWTOT(IROOM,IPROD+2,UU)
            ELSE IF (HINTER.GE.HCEIL.AND.FLWTOT(IROOM,IP,UU).LT.XX0)
     +          THEN
              XPRIME(IPRODU) = FLWTOT(IROOM,IPROD+2,UU)
            ELSE
              XPRIME(IPRODU) = XX0
            END IF
            IF (HINTER.GT.XX0) THEN
              XPRIME(IPRODL) = FLWTOT(IROOM,IPROD+2,LL)
            ELSE IF (HINTER.LE.XX0.AND.FLWTOT(IROOM,IP,LL).GT.XX0) THEN
              XPRIME(IPRODL) = FLWTOT(IROOM,IPROD+2,LL)
            ELSE
              XPRIME(IPRODL) = XX0
            END IF
 90       CONTINUE
 100    CONTINUE
      END IF
```

Finally, the residuals are calculated. These are simply the difference between the solution vector calculated by RESID and that passed to RESID by the differential equation solver. This is done in several parts to correspond with the layout of the solution vector.

```
C
C     RESIDUALS FOR PRESSURE, LAYER VOLUME, AND LAYER TEMPERATURES
C
      DO 110 I = NOFP + 1, NOFP + NM1
        DELTA(I) = XPRIME(I) - XPSOLVE(I)
 110  CONTINUE
      DO 120 I = NOFTU + 1, NOFTU + 3 * NM1
        DELTA(I) = XPRIME(I) - XPSOLVE(I)
 120  CONTINUE
C
C     CONDUCTION RESIDUAL
C
      CALL CNHEAT(UPDATE,DT,FLXTOT,DELTA)
C
C     RESIDUALS FOR GAS LAYER SPECIES, NOTE THAT DASSL IS NOT SOLVING FOR
C     SPECIES NOW, THIS IS DONE IN SOLVE
C
      DO 130 I = NOFPRD + 1, NOFPRD + 2 * NPROD * NM1
        DELTA(I) = XPRIME(I) - XPSOLVE(I)
 130  CONTINUE
C
C     RESIDUAL FOR HVAC SPECIES
C
      IF (NPROD.NE.0) THEN
        DO 140 I = NOFHVPR + 1, NOFHVPR + NLSPCT * NHVSYS
          DELTA(I) = XPRIME(I) - XPSOLVE(I)
 140    CONTINUE
      END IF
      IF (IPAR(2).EQ.SOME) THEN
        NPROD = NPRODSV
      END IF
      RETURN
      END
```

4.4 Interface to the CFAST Physical Interface Routines

Each physical interface routine calculates flow and/or flux terms as appropriate for all rooms and/or surfaces of the simulation being modeled. These flow and flux terms are the effect of the phenomenon on each of the layers and/or surfaces and includes flows due to mass, enthalpy and products of combustion. Rather than using multiple variables for each room, these are organized into a single array for each phenomenon. This structure is shown in Figure 23. To illustrate the organization of the physical interface routines, the following outlines the steps in calculating one of the phenomena.

The physical interface routine, FIRES, calculates the rates of addition of mass, enthalpy, and species into all layers in all rooms from all fires in a simulation. For each fire, the following scheme is employed:

1. Initialize the fire data structure, FLWF, to zero.

2. For each specified fire, the routine PYROLS (for the main fire) or OBJINT (for object fires) calculates time dependent quantities for the time of interest by interpolating between the time points specified by the user. The routine DOFIRE calculates the plume entrainment rate.

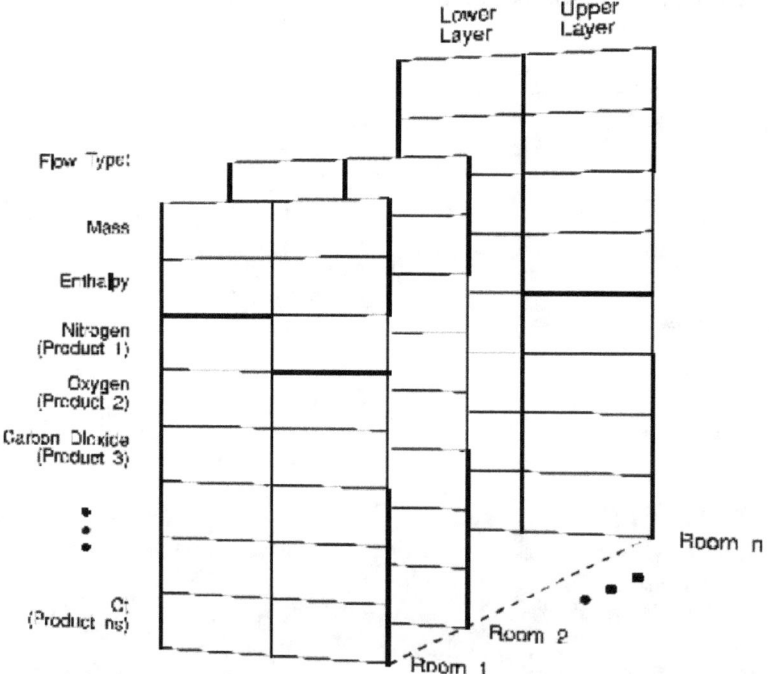

Figure 23. Data structure for flow and/or flux terms returned
from physical interface routines to the control routine RESID.

3. For a type 1 (unconstrained) fire, the routine DOFIRE sets the burning rate to the pyrolysis rate. The heat release rate is found by multiplying the burning rate by the heat of combustion.

4. For a type 2 (constrained) fire, the prescribed chemistry scheme, discussed above, is used to constrain the burning rate based on *both* the fuel and oxygen available. This chemistry scheme is implemented in the routine CHEMIE. This calculation is done for both the lower layer (from the mass entrained by the plume) and for burning in the upper layer (with oxygen and fuel available in the layer).

5. Sum the contributions from all fires into the fire data structure for return to the control routine. The following code fragment is typical of those in all of the physical interface routines:

```
      FLWF(LFBO,M,UPPER) = FLWF(LFBO,M,UPPER) + EMS(LFBO)
      FLWF(LFBO,M,LOWER) = FLWF(LFBO,M,LOWER) - EME(LFBO)
      FLWF(LFBO,Q,UPPER) = FLWF(LFBO,Q,UPPER) + QF(LFBO) + QEME + QEMP
      FLWF(LFBO,Q,LOWER) = FLWF(LFBO,Q,LOWER)          - QEME
      QLPQUF(LFBO)       = QLPQUF(LFBO)        + QF(LFBO)        + QEMP
      DO 40 LSP = 1, NS
        FLWF(LFBO,LSP+2,UPPER) = FLWF(LFBO,LSP+2,UPPER) +
     +     XNIMS(UPPER,LSP)
        FLWF(LFBO,LSP+2,LOWER) = FLWF(LFBO,LSP+2,LOWER) +
     +     XNIMS(LOWER,LSP)
 40   CONTINUE
```

5 Discussion and Summary

The software consists of a collection of data and computer programs which are used to *simulate* the important time-dependent phenomena involved in fires. The major functions provided include calculation of:

- the buoyancy-driven as well as forced transport of this energy and mass through a series of specified compartments and connections (e.g., doors, windows, cracks, ducts),

- the resulting temperatures, smoke optical densities, and gas concentrations after accounting for heat transfer to surfaces and dilution by mixing with clean air.

As can be seen from this list, fire modeling involves an interdisciplinary consideration of physics, chemistry, fluid mechanics, and heat transfer. In some areas, fundamental laws (conservation of mass, energy, and momentum) can be used, whereas in others empirical correlations or even "educated guesses" must be employed to bridge gaps in existing knowledge. The necessary approximations required by operational practicality result in the introduction of uncertainties in the results. The user should understand the inherent assumptions and limitations of the programs, and use these programs judiciously – including sensitivity analyses for the ranges of values for key parameters – in order to make estimates of these uncertainties. This section provides an overview of these assumptions and limitations.

5.1 Specified Fire Limitations

An important limitation of CFAST is the absence of a fire growth model. At the present time, it is not practical to adapt currently available fire growth models for direct inclusion in CFAST. CFAST utilizes a user specified fire expressed in terms of an energy and mass release rates of the burning item(s) at various times. Such data can be obtained by measurements taken in large- and small-scale calorimeters, or from compartment burns. Examples of their associated limitations are as follows:

1. For a large-scale calorimeter, a product (e.g., chair, table, bookcase) is placed under a large collection hood and ignited by a 50 kW gas burner (simulating a wastebasket) placed adjacent to the item for 120 s. The combustion process then proceeds under assumed "free-burning" conditions, and the release rate data are measured. Potential sources of uncertainty include measurement errors related to the instrumentation and the degree to which "free-burning" conditions are not achieved (e.g., radiation from the gases under the hood or from the hood itself, and restrictions in the air entrained by the object causing locally reduced oxygen concentrations affecting the combustion chemistry). There are limited experimental data for upholstered furniture which suggest that prior to the onset of flashover in a compartment, the

influence of the compartment on the burning behavior of the item is small. The differences obtained from the use of different types or locations of ignition sources have not been explored. These factors are discussed in reference [68].

2.   Where small-scale calorimeter data are used, procedures are available to extrapolate to the behavior of a full-size item. These procedures are based on empirical correlations of data which exhibit significant scatter, thus limiting their accuracy. For example, for upholstered furniture, the peak heat release rates estimated by the "triangular approximation" method averaged 91 % (range 46 to 103 %) of values measured for a group of 26 chairs with noncombustible frames, but only 63 per cent (range 46 to 83 %) of values measured for a group of 11 chairs with combustible frames [69]. Also, the triangle neglects the "tails" of the curve; these are the initial time from ignition to significant burning of the item, and the region of burning of the combustible frame, after the fabric and filler are consumed.

3.   The provided data and procedures only relate directly to burning of items initiated by relatively large flaming sources. Little data are currently available for release rates under smoldering combustion, or for the high external flux and low oxygen conditions characteristic of post-flashover burning. While the model allows multiple items burning simultaneously, it does not account for the synergy of such multiple fires. Thus, for other ignition scenarios, multiple items burning simultaneously (which exchange energy by radiation and convection), combustible interior finish, and post-flashover conditions, the move can give estimates which are often nonconservative (the actual release rates would be *greater* than estimated). At present, the only sure way to account for all of these complex phenomena is to conduct a full-scale compartment burn and use the pyrolysis rates directly.

5.2 Zone Model and Transport Limitations

The basic assumption of all zone fire models is that each compartment can be divided into a small number of control volumes, each of which is internally uniform in temperature and composition. In CFAST all compartments have two zones except for the fire room which has an additional zone for the plume.

The boundary between the two layers in a compartment is called the interface. It has generally been observed that buoyantly stratified layers form in the spaces close to the fire. While in an experiment the temperature can be seen to vary within a given layer, these variations are small compared to the temperature difference between the layers.

Beyond the basic zone assumptions, the model typically involves a mixture of established theory (e.g., conservation equations), empirical correlations where there are data but no theory (e.g., flow and entrainment coefficients), and approximations where there are neither (e.g., post-flashover combustion chemistry) or where their effect is considered secondary compared to the "cost" of inclusion. An example of a widely used assumption is that the estimated error from ignoring the variation of the thermal properties of structural materials with temperature is small. While this information would be fairly simple to add to the computer code, data are scarce over a broad range of temperatures even for the most common materials.

The user should be aware of the general limits of zone modeling and some specific manifestations in CFAST. These include the following:

1.   Burning can be constrained by the available oxygen. However, this "constrained fire" (a "type 2" fire, see page 17) is not subject to the influences of radiation to enhance its burning rate, but is influenced by the oxygen available in the compartment. If a large mass loss rate is entered, the model will follow this input until there is insufficient oxygen available for that quantity of fuel to burn in the compartment. The unburned fuel (sometimes called excess pyrolyzate) is tracked as it flows out in the door jet, where it can entrain more oxygen. If this mixture is within the user-specified flammable range, it burns in the door plume. If not, it will be tracked throughout the building until it eventually collects as unburned fuel or burns in a vent. The enthalpy released in the fire compartment and in each vent, as well as the total enthalpy released, is detailed in the output of the model. Since mass and enthalpy are conserved, the total will be correct. However, since combustion did not take place adjacent to the burning object, the actual mass burned could be lower than that specified by the user. The difference will be the unburned fuel.

2.   An oxygen combustion chemistry scheme is employed only in constrained (type 2) fires. Here user-specified hydrocarbon ratios and species yields are used by the model to predict concentrations. A balance among hydrogen, carbon, and oxygen molecules is maintained. Under some conditions, low oxygen can change the combustion chemistry, with a resulting increase in the yields of products of incomplete combustion such as CO. Guidance is provided on how to adjust the CO/CO_2 ratio. However, not enough is known about these chemical processes to build this relationship into the model at the present time. Some data exist in reports of full-scale experiments (e.g., reference [70]) which can assist in making such determinations.

3.   The entrainment coefficients are empirically determined values. Small errors in these values will have a small effect on the fire plume or the flow in the plume of gases exiting the door of that compartment. In a multi-compartment model such as CFAST, however, small errors in each door plume are multiplicative as the flow proceeds through many compartments, possibly resulting in a significant error in the furthest compartments. The data available from validation experiments [71] indicate that the values for entrainment coefficients currently used in most zone models produce good agreement for a three-compartment configuration. More data are needed for larger numbers of compartments to study this further.

4.   In real fires, smoke and gases are introduced into the lower layer of each compartment primarily due to mixing at connections between compartments and from the downward flows along walls (where contact with the wall cools the gas and reduces its buoyancy). Doorway mixing has been included in CFAST, using the same empirically derived mixing coefficients as used for calculating fire plume entrainment. Downward wall flow has not been included. This could result in underestimates of lower layer temperatures and species concentration.

5.3 Model Evaluation

The ASTM guide for evaluating the predictive capability of fire models [72] identifies four areas important for fire model evaluation: 1) model and scenario definition, 2) theoretical basis and assumptions in the model, 3) mathematical and numerical robustness of the model, and 4) quantifying the uncertainty and accuracy of the model. The first two of these are largely documentation issues for the model developer. For the CFAST model, a user's guide is available to guide model and scenario definition [73]. This publication provides details of the theoretical basis and assumptions in the CFAST model. Additional guidance is available in the ASTM guide for fire model documentation [74]. The work of Forney [75] examines the numerical robustness of fire models using the CFAST model as an example. Sensitivity analysis and experimental comparisons are the primary focus of the final area. An overview of published literature related to FAST / CFAST in these areas is presented below.

A number of researchers have studied the level of agreement between computer fire models and real-scale fires. These comparisons fall into two broad categories: fire reconstruction and comparison with laboratory experiments. Both categories provide a level of verification for the models used. Fire reconstruction, although often more qualitative, provides a higher degree of confidence for the user when the models successfully simulate real-life conditions. Comparisons with laboratory experiments, however, can yield detailed comparisons that can point out weaknesses in the individual phenomena included in the models. The comparisons made to date are mostly qualitative in nature. The level of agreement between the models and experiment is typically reported as "favorable," "satisfactory," "well predicted," "successful," or "reasonable." Some of the comparisons in the literature are reviewed below.

Nelson [76] used simple computer fire models along with existing experimental data to develop an analysis of a large high-rise building fire. This analysis showed the value of available analytical calculations in reconstructing the events involved in a multiple-story fire. Bukowski [77], [78], [79] has applied the FAST and CFAST models in several fatal fire reconstructions. Details of the fires including temperatures, vent flows, and gas concentrations were consistent with observed conditions and witness accounts.

Several studies comparing model predictions with experimental measurements are available. Deal [80] reviewed four computer fire models (CCFM, FIRST, FPETOOL[81] and FAST) to ascertain the relative performance of the models in simulating fire experiments in a small room. All the models simulated the experimental conditions including temperature, species generation, and vent flows, "quite satisfactorily." Duong [82] studied the predictions of several computer fire models (CCFM, FAST, FIRST, and BRI), comparing the models with one another and with large fires in an aircraft hanger. For a 4 MW fire size, he concluded that all the models are "reasonably accurate." At 36 MW, however, "none of the models did well." Beard [83], [84] evaluated four fire models (ASET, FAST, FIRST, and JASMINE[85]) by modeling three well-documented experimental fires, ranging in scope from single compartments to a large-department-store space with closed doors and windows. He provides both a qualitative and quantitative assessment of the models ability to predict temperature, smoke obscuration, CO concentration, and layer interface position (for the zone-based models). Peacock, Jones, and Bukowski [86] and Peacock, et. al. [87] compared the CFAST model to a range of experimental fires. The model provided predictions of the magnitude and trends (time to critical conditions and general curve shape) for the experiments studied which range in quality from within a few % to a factor of two or three of the measured values.

The CFAST model has been subjected to extensive evaluation studies by NIST and others. Although differences between the model and the experiments were evident in the studies, they are typically explained by limitations of the model and of the experiments. Like all predictive models, the best predictions come with a clear understanding of the limitations of the model and of the inputs provided to the calculations.

6 References

[1] Friedman, R., "Survey of Computer Models for Fire and Smoke," *Factory Mutual Research Corp.*, Norwood, MA, 02062 1990.

[2] Cooper, L.Y., "A Mathematical Model for Estimating Available Safe Egress Time in Fires," *Fire and Materials.* 1982, *6(4)*, 135-144.

[3] Babrauskas, V., "COMPF2-A Program for Calculating Post-Flashover Fire Temperatures," *Natl. Bur. Stand. (U.S.)* 1979, *Tech. Note 991*, 76 p.

[4] Davis, W. D. and Cooper, L. Y., "Computer Model for Estimating the Response of Sprinkler Links to Compartment Fires With Draft Curtains and Fusible Link-Actuated Ceiling Vents," *Fire Technology* 1991, *27 (2)*, 113-127.

[5] Tanaka, T., "A Model of Multiroom Fire Spread," *Nat. Bur. Stand. (U.S.)* 1983, *NBSIR 83-2718*, 175 p.

[6] Jones, W. W., A Multicompartment Model for the Spread of Fire, Smoke and Toxic Gases, Fire Safety Journal 9, 55 (1985); Jones, W. W. and Peacock, R. D., Refinement and Experimental Verification of a Model for Fire Growth and Smoke Transport, Proceedings of the 2nd International Symposium on Fire Safety Science, Tokyo (1989); Jones, W. W. and Peacock, R. D., "Technical Reference Guide for FAST Version 18" Natl. Inst. Stand. Techol. Tech. Note 1262 (1989).

[7] Forney, G. P. and Cooper, L. Y., The Consolidated Compartment Fire Model (CCFM) Computer Application CCFM.VENTS - Part II: Software Reference Guide, Nat. Inst. Stand. Technol., NISTIR 90-4343 (1990).

[8] Jones, W. W. and Forney, G. P. "A Programmer's Reference Manual for CFAST, the Unified Model of Fire Growth and Smoke Transport," *Natl. Inst. Stand. Technol.* 1990, *Tech. Note 1283*, 104 p.

[9] Mitler, H. E. "Comparison of Several Compartment Fire Models: An Interim Report," *Natl. Bur. Stand. (U.S.)* 1985, *NBSIR 85-3233*, 33 p.

[10] Jones, W. W. "A Review of Compartment Fire Models," *Natl. Bur. Stand. (U.S.)* 1983, *NBSIR 83-2684*, 41 p.

[11]   Cooper L. Y. and Forney, G. P., The consolidated compartment fire model (CCFM) computer application CCFM-VENTS – part I: Physical reference guide. Natl. Inst. Stand. Technol., NISTIR 4342 (1990).

[12]   Forney, G. P. and Moss, W. F., Analyzing and Exploiting Numerical Characteristics of Zone Fire Models, Natl. Inst. Stand. Technol., NISTIR 4763 (March 1992).

[13]   Jones, W. W. and Bodart, X., Buoyancy Driven Flow as the Forcing Function of Smoke Transport Models, Natl. Bur. Stand. (U. S.), NBSIR 86-3329 (1986).

[14]   Rehm, R. G and Forney, G. P., "A Note on the Pressure Equations Used in Zone Fire Modeling," Natl. Inst. Stand. Technol., NISTIR 4906 (1992).

[15]   Babrauskas, V., "Development of the Cone Calorimeter - A Bench Scale Heat Release Rate Apparatus Based on Oxygen Consumption," *Fire and Materials 8,* 1984, p 81.

[16]   Thornton, "The Relation of Oxygen to the Heat of Combustion of Organic Compounds," Philosophical Magazine and J. of Science, **33** (1917).

[17]   Huggett, C., "Estimation of the Rate of heat Release by Means of Oxygen Consumption," J. of Fire and Flammability, **12**, pp. 61-65 (1980).

[18]   Standard Test Method for Heat and Visible Smoke Release for Materials and Products Using and Oxygen Consumption Calorimeter, ASTM E1354-90, *American Society for Testing and Materials,* Philadelphia, PA 1990.

[19]   Standard Test Method for Heat and Visible Smoke Release for Materials and Products Using and Oxygen Consumption Calorimeter, ASTM E1354-90, American Society for Testing and Materials, Philadelphia, PA (1990).

[20]   Morehart, J. H., Zukowski, E. E. and Kubota, T., Characteristics of Large Diffusion Flames Burning in a Vitiated Atmosphere, Third International Symposium on Fire Safety Science, Edinburgh (1991).

[21]   McCaffrey, B. J., "Momentum Implications for Buoyant Diffusion Flames," *Combustion and Flame 52,* 1983, p. 149.

[22]   Cetegen, B. M., "Entrainment and Flame Geometry of Fire Plumes," Ph.D. Thesis, California Institute of Technology, Pasadena 1982.

[23]   Drysdale, D., "An Introduction to Fire Dynamics," John Wiley and Sons, New York, 143 p. (1985).

[24]   Tewarson, A., Combustion of Methanol in a Horizontal Pool Configuration, Factory Mutual Research Corp., Norwood, MA, Report No. RC78-TP-55 (1978).

[25]   McCaffrey, B. J., Entrainment and Heat Flux of Buoyant Diffusion Flames, Natl. Bur. Stand. (U. S.), NBSIR 82-2473, 35 p. (1982).

[26]   Koseki, H., Combustion Properties of Large Liquid Pool Fires, Fire Technology, **25(3)**, 241-255 (1989).

[27]   Quintiere, J. G, Steckler, K., and Corley, D., An Assessment of Fire Induced Flows in Compartments, Fire Science and Technology *4*, 1 (1984).

[28]   Quintiere, J. G, Steckler, K. and McCaffrey, B. J., "A Model to Predict the Conditions in a Room Subject to Crib Fires," First Specialist Meeting (International) of the Combustion Institute, Talence, France 1981.

[29]   Cooper, L. Y., Calculation of the Flow Through a Horizontal Ceiling/Floor Vent, Natl. Inst. Stand. Technol., NISTIR 89-4052 (1989).

[30]   Klote, J.K. and Milke, J.A., Design of Smoke Management Systems, American Society of Heating, Refrigerating and Air-conditioning Engineers, Atlanta, GA (1992).

[31]   ASHRAE Handbook HVAC Systems and Equipment, American Society of Heating, Refrigerating and Air-Conditioning Engineers, Atlanta, GA (1992).

[32]   Klote, J. H., A Computer Model of Smoke Movement by Air Conditioning Systems, NBSIR 87-3657 (1987).

[33]   1989 ASHRAE Handbook Fundamentals, American Society of Heating, Refrigeration and Air Condition Engineers, Inc., Atlanta, GA 1989.

[34]   Jorgensen, R. 1983. Fan Engineering, Buffalo Forge Co., Buffalo, NY.

[35]   ASHRAE. Handbook of Fundamentals, Chapter 32 Duct Design, American Society of Heating, Refrigerating and Air-Conditioning Engineers, Atlanta, GA (1993).

[36]   Murdock, J.W., Mechanics of Fluids, Marks' Standard Handbook for Mechanical Engineers, 8th ed., Baumeister, *et al.* editors, McGraw, New York, NY (1978).

[37]   Schlichting, H., Boundary Layer Theory, 4th ed., Kestin, J. Translator, McGraw, New York, NY (1960).

[38]   Huebscher, R.G, Friction Equivalents for Round, Square and Rectangular Ducts, ASHVE Transactions (renamed ASHRAE Transactions), Vol. 54, pp 101-144 (1948).

[39]   Heyt, J.W. and Diaz, J. M., Pressure Drop in Flat-Oval Spiral Air Duct, ASHRAE Transactions, Vol. 81, Part 2, p 221-230 (1975).

[40]   Colebrook, C.F., Turbulent Flow in Pipes, With Particular Reference to the Transition Region Between the Smooth and Rough Pipe Laws, Journal of Institution of Civil Engineers (London, England), Vol 11, pp 133-156 (1938-1939).

[41]   Moody, L.F., Friction Factors for Pipe Flow, Transactions of ASME, Vol 66, p 671-684 (1944).

[42]   McGrattan, K. B., Baum, H. R., and Rehm, R. G, Large Eddy Simulations of Smoke Movement, Fire Safety J. **30** (2), pp 161-178 (1998).

[43]   Forney, G P., Computing Radiative Heat Transfer Occurring in a Zone Fire Model, Natl. Inst. Stand. Technol., NISTIR 4709 (1991).

[44]   Siegel, R. and Howell, J. R., Thermal Radiation Heat Transfer, Hemisphere Publishing Corporation, New York, second ed. (1981).

[45]   Hottel, H. C., Heat Transmission, McGraw-Hill Book Company, New York, third ed. (1954).

[46]   Hottel, H. and Cohen, E., Radiant Heat Exchange in a Gas Filled Enclosure: Allowance for non-uniformity of Gas Temperature, American Institute of Chemical Engineering Journal *4*, 3 (1958).

[47]   Yamada, T. and Cooper, L. Y., Algorithms for Calculating Radiative Heat Exchange Between the Surfaces of an Enclosure, the Smoke Layers and a Fire, Building and Fire Research Laboratory Research Colloquium, July 20, 1990.

[48]   Jones, W. W. and Forney, G P., Improvement in Predicting Smoke Movement in Compartmented Structures, Fire Safety J., **21**, pp 269-297 (1993).

[49]   Schlichting, H., "Boundary Layer Theory," translated by J. Kestin, Pergammon Press, New York, 1955.

[50]   Atreya, A., "Convection Heat Transfer," Chapter 1-4 in SFPE Handbook of Fire Protection Engineering, DiNenno, P. J, Beyler, C. L., Custer, R. L. P., Walton, W. D., and Watts, J. M., eds., National Fire Protection Association, Quincy, MA (1988).

[51]   Golub, G H. and Ortega, J. M., Scientific Computing and Differential Equations, An Introduction to Numerical Methods. Academic Press, New York (1989).

[52]   Strang, G and Fix, G J., An Analysis of the Finite Element Method. Prentice-Hall, Englewood Cliffs, New Jersey (1973).

[53]   Moss, W. F. and Forney, G. P., "Implicitly coupling heat conduction into a zone fire model." Natl. Inst. Stand. Technol., NISTIR 4886 (1992).

[54]   Robert Siegel and John R. Howell. Thermal Radiation Heat Transfer. Hemisphere Publishing Corporation, New York, second edition, 1981.

[55]   Glenn P. Forney. Computing radiative heat transfer occurring in a zone fire model. Internal Report 4709, National Institute of Standards and Technology, 1991.

[56]   Cooper, L. Y., Fire-Plume-Generated Ceiling Jet Characteristics and Convective Heat Transfer to Ceiling and Wall Surfaces in a Two-Layer Zone-Type Fire Environment, Natl. Inst. Stand. Technol., NISTIR 4705, 57 p. (1991).

[57]   Cooper, L. Y., Heat Transfer in Compartment Fires Near Regions of Ceiling-Jet Impingement on a Wall. J. Heat Trans., 111, pp. 455-460 (1990).

[58]   Cooper, L. Y., Ceiling Jet-Driven Wall Flows in Compartment Fires. Combustion Sci. and Technol., 62, pp. 285-296 (1988).

[59]   Jaluria, Y. and Cooper, L. Y., Negatively Buoyant Wall Flows Generated in Enclosure Fires. Progress in Energy and Combustion Science, 15, pp. 159-182 (1989).

[60]   Leonard Y. Cooper. Fire-plume-generated ceiling jet characteristics and convective heat transfer to ceiling and wall surfaces in a two-layer zone-type fire environment: Uniform temperature and walls. Technical Report 4705, National Institute of Standards and Technology, 1991.

[61]   Gunnar Heskestad and Herbert F. Smith. Investigation of a new sprinkler sensitivity approval test: The plunge test. Technical Report Serial No. 22485 2937, Factory Mutual Research Corporation, Norwood, MA, 1976. RC 76-T-50.

[62]   D. Madrzykowski and R.L. Vettori. A sprinkler fire suppression algorithm for the gsa engineering fire assessment system. Technical Report 4833, National Institute of Standards and Technology, 1992.

[63]   David D. Evans. Sprinkler fire suppression for hazard. Technical Report 5254, National Institute of Standards and Technology, 1993.

[64]   Seader, J., and Einhorn, I., "Some Physical, Chemical, Toxicological and Physiological Aspects of Fire Smokes,: *Sixteenth Symposium (International) on Combustion.* The Combustion Institute, Pittsburgh, PA 1976, pp. 1423-1445.

[65] Galloway, F. M., Hirschler, M. M., "A Model for the Spontaneous Removal of Air-borne Hydrogen Chloride by Common Surfaces," *Fire Safety Journal 14,* 1989, pp. 251-268.

[66] Galloway, F. M., Hirschler, M. M., "Transport and Decay of Hydrogen Chloride: Use of a Model to Predict Hydrogen Chloride Concentrations in Fires Involving a Room-Corridor-Room Arrangement," *Fire Safety Journal 16,* 1990, pp. 33-52.

[67] Brenan, K. E., Campbell, S. L., and Petzold, L. R., Numerical Solution of Initial-Value Problems in Differential-Algebraic Equations, Elsevier Science Publishing, New York (1989).

[68] Babrauskas, V., Lawson, J. R., Walton, W. D., Twilley, W. H., "Upholstered Furniture Heat Release Rates Measured with a Furniture Calorimeter," Natl. Bur. Stand. (U.S.), NBSIR 82-2604, 1982.

[69] Babrauskas, V. and Krasny, J. F., "Fire Behavior of Upholstered Furniture," Natl. Bur. Stand. (U.S.), Monogr. 173, 1985.

[70] Lee, B.T., "Effect of Ventilation on the Rates of Heat, Smoke, and Carbon Monoxide Production in a Typical Jail Cell Fire," Natl. Bur. Stand. (U.S.), NBSIR 82-2469, 1982.

[71] Peacock, R. D., Davis, S., Lee, B. T., "An Experimental Data Set for the Accuracy Assessment of Room Fire Models," Natl. Bur. Stand. (U.S.), NBSIR 88-3752, April 1988, p. 120.

[72] Standard Guide for Evaluating the Predictive Capability of Fire Models, ASTM E 1355, Annual Book of ASTM Standards, Vol. 04.07, American Society for Testing and Materials, Philadelphia (1995).

[73] Peacock, R. D., Reneke, P. A., Jones, W. W., Bukowski, R. W., and Forney, G P., "A User's Guide for FAST, Engineering Tools for Estimating Fire Growth and Smoke Transport," Natl. Inst. Stand. Technol., Special Pub. 921, 180 pp. 1997.

[74] Standard Guide for Guide for Documenting Computer Software for Fire Models, ASTM E 1472, Annual Book of ASTM Standards, Vol. 04.07, American Society for Testing and Materials, Philadelphia (1995).

[75] Forney, G P., and Moss, W. F., " Analyzing and Exploiting Numerical Characteristics of Zone Fire Models," *Fire Science and Technology,* **14**, No. 1/2, 49-60, 1994.

[76] Nelson, H. E., "An Engineering View of the Fire of May 4, 1988 in the First Interstate Bank Building, Los Angeles, California," *Natl. Inst. Stand. Technol. NISTIR 89-4061*, 39 p (1989).

[77]   Bukowski, R.W., "Reconstruction of a Fatal Residential Fire at Ft. Hood, Texas," *First HAZARD I Users' Conference,* National Institute of Standards and Technology, Gaithersburg, MD, June 5-6 (1990).

[78]   Bukowski, R. W., "Analysis of the Happyland Social Club Fire With HAZARD I," Fire and Arson Investigator, Vol. 42, No. 3, pp.36-47 (1992).

[79]   Bukowski, R. W., "Modeling a Backdraft: The Fire at 62 Watts Street," NFPA Journal, Vol. 89, No. 6, pp. 85-89 (1995).

[80]   Deal, S. "A Review of Four Compartment Fires with Four Compartment Fire Models," *Fire Safety Developments and Testing*, Proceedings of the Annual Meeting of the Fire Retardant Chemicals Association. October 21-24, 1990, *Ponte Verde Beach, Florida*, 33-51.

[81]   Nelson, H. E., "FPETOOL: Fire Protection Engineering Tools for Hazard Estimation," *Natl. Inst. Stand. Technol.* 1990, *NISTIR 4380*, 120 p.

[82]   Duong, D. Q., "The Accuracy of Computer Fire Models: Some Comparisons with Experimental Data from Australia," *Fire Safety J.* 1990, *16(6)*, 415-431.

[83]   Beard, A., Evaluation of Fire Models: Part I – Introduction. Fire Safety J. **19**, 295-306 (1992).

[84]   Beard, A., "Evaluation of Fire Models: Overview," Unit of Fire Safety Engineering, University of Edinburgh, Edinburgh, UK 1990.

[85]   Cox, G and Kumar, S., "Field modeling of fire in forced ventilated enclosures," *Combust. SCi. Technol.*, **52** 7-23 1987.

[86]   Peacock, R. D., Jones, W. W., and Bukowski, R. W., Verification of a Model of Fire and Smoke Transport, Fire Safety J. **21**, 89-129 (1993).

[87]   Peacock, R. D.; Reneke, P. A.; Forney, C. L.; and Kostreva, M. M, "Issues in Evaluation of Complex Fire Models," Fire Safety Journal, **30**, 103-136, 1998.

7 Acknowledgments

This work was partially funded by the Office of Naval Research, Code 334, under the Damage Control Task of the Surface Ship Hull, Mechanical and Electrical Technology Program (PE0602121N).

Appendix A A Note on Total Gas Layer Absorption

John B. Hoover and Jean L. Bailey
Code 6180
US Naval Research Laboratory
Washington, DC 20375

A.1 Theoretical Considerations

Absorption is known to be a function of several parameters, including the concentrations of various gases (primarily particle size distribution of the soot, the temperature of the radiation source and the temperature of the absorbing gas relationship is indicated by

$$A = \text{function } ([CO_2], [H_2O], \text{soot}, T, ...) \tag{A1}$$

Given a volume filled with gases and suspended particulates, radiation traversing that volume will be transmitted, ab:

$$1 = T + A + S \tag{A2}$$

where T, A, and S represent the energy fractions which are transmitted, absorbed and scattered, respectively. For ou component, because: (a) scattering would add considerable complexity (computational time) to the model; and (b) m detailed knowledge of the particulate size distribution and of the optical properties of individual particles.

As a reasonable approximation, we assume exponential attenuation, so transmittance is

$$T = \exp(-\alpha C L) \tag{A3}$$

where α is the specific absorption coefficient, C is the concentration of the absorbing species and L is the path length dimensionless, α is in units of $(CL)^{-1}$.

In general, the specific absorption coefficient is a function of wavelength, so we should calculate total transmittance b transmittance over the entire wavelength region of interest. Because that calculation is computationally expensive, w average value of α is valid for the entire wavelength region. This is a reason-able approximation for black- or gray b continuous absorption spectra. We also assume that absorption is only due to soot, H_2O and CO_2.

Applying eq (A3) to soot, the transmittance becomes

$$T_S = \exp(-\alpha_S C_S L) \tag{A4}$$

where α_S is the effective soot specific absorption coefficient and C_S is the soot concentration.

Making the assumption that absorption is additive, except for a correction for band overlap, the gas absorbance will

$$A_G = A_{H2O} + A_{CO2} - C \qquad (A5)$$

For typical fire conditions, the overlap amounts to about half of the CO_2 absorbance [A1] so the gas transmittance is

$$T_G = 1 - A_{H2O} - 0.5\ ACO2 \qquad (A6)$$

Due to the band structure of gas absorption spectra, the assumption of continuous absorption is not a good approxim were estimated from graphs of detailed band absorbance calculations [A2].

The total transmittance of a gas-soot mixture is the product of the gas and soot transmittances

$$T_T = T_S T_G \qquad (A7)$$

Substitution of eq (A4) and (A6) into eq (A7) yields

$$T_T = \exp(-\alpha_S C_S L)\ [1 - A_{H2O} - 0.5\ ACO2] \qquad (A8)$$

The parameters α_S and C_S may be lumped together as κ, which, in the optically thin limit, is the Planck mean absorpt limit, is the Rosseland mean absorption coefficient. For the entire range of optical thicknesses, Tien, *et. al.* [A3] repo

$$\kappa = k\ f_V\ T \qquad (A9)$$

where k is a constant which depends on the optical properties of the soot particles, f_V is the soot volume fraction an than attempt to calculate κ from first principals, we have taken a semi-empirical approach, using values of κ, f_V, and and Tien [A4] for various fuel types. As seen in C-1, κ is quite variable [mean: 14.2; standard deviation: 13.5 (95%) after normalizing for f_V and T, k is found to be nearly constant [mean: 1195.5; standard deviation: 43.6 (4%)]. There valid approximation for a wide range of fuels. The soot volume fraction, f_V, is calculated from the soot mass, soot der assumed to be in thermal equilibrium with the gas layer.

Equation (A8) was developed for the case of point-to-point transfer, in which the value of L is known. In the contex ambiguous since the source is distributed over the volume of a layer and the destination is the surface area of the laye length must be calculated. The mean beam length concept treats an emitting volume as if it were a hemisphere of a rad center of the circular base is equal to the average boundary flux produced by the real volume. The value of this radiu but, for an arbitrary shape, Tien *et. al.* [A5] have found an approximate value to be

$$L = c\ 4\ V/A \qquad (A10)$$

110

where L is the mean beam length in meters, c is a constant (approximately 0.9, for typical geometries), V is the emitti
surface area (meter^2) of the gas volume. The volume and surface area are calculated from the dimensions of the laye
first term of eq (A8).

Edwards' [A2] absorbance data for H_2O and CO_2 are reported as log(emissivity) versus log(gas temperature), with
temperature is in Kelvin and gas concentrations is expressed a pressure-path length, in units of atmosphere-meters.
into a look-up table, implemented as a two-dimensional array of log(emissivity) values, with indices based on tempe

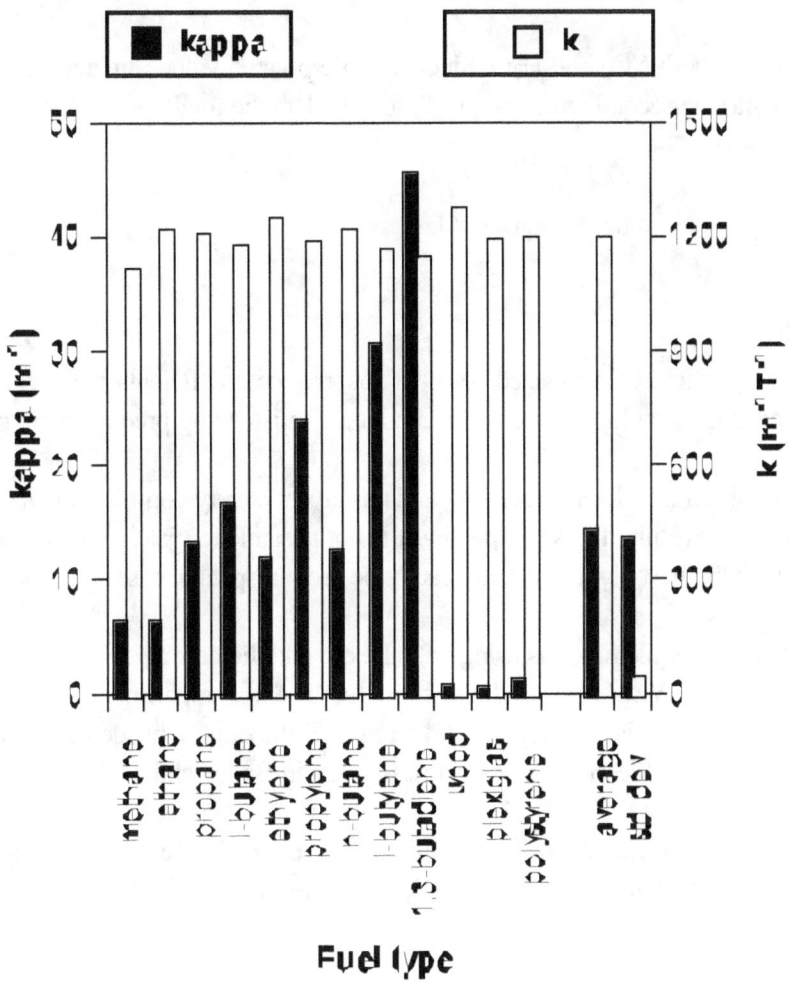

Figure A-1. Comparison of variability of κ and k for various fuels.

It should be noted that the literature usually reports the results of emission, rather than absorption, in experiments and
radiation impinging on gray bodies which are in thermal equilibrium with their surroundings, absorption and emission
assumptions for the case of soot particles suspended in air, therefore the use of emission data to estimate absorbance
approximation is not as good in the case of gases, but the alternative, a detailed absorbance band calculation, is comp
have assumed that absorbance and emissivity are equivalent.

111

The number of moles of each gas is found using the gas molecular weights and CFAST's estimated gas mass fraction in units of atmosphere-meters, are calculated from the number of moles, the universal gas constant, temperature, laye (A10). For each gas, the log(absorbance) is estimated from the corresponding look-up table by linear interpolation in log(concentration) domains. In the event that the required absorbance lies outside the temperature or concentration r acceptable value is returned. Error flags are also returned, indicating whether each parameter was in or out of range a low. Presently, these flags are only used to generate a warning message (sent to the default output device, unit 6) in t the future, they could be used for other purposes. For example, they might trigger a more complex extrapolation rou

This entire process is carried out for both CO_2 and H_2O. Since the interpolated values are actually log(absorbance), gas absorbances are substituted into the second term of eq (A8) to calculate the total transmittance. Finally, the latter coefficient

$$\alpha_{EFF} = -\ln(T_T)/L \qquad\qquad (A11)$$

A.2 Results

Prior to the development of this algorithm, CFAST used constant absorbance coefficients of 0.5 for upper layers and for the case of sooty upper layers and clean lower layers, but can lead to misleading predictions in other cases.

The interaction of radiative transport with walls, floor, ceiling and gas layers is very complex and the addition of varia situation even more complicated. As a result of this complexity, it is not possible to make general statements regardin coefficient. However, the effects of the new algorithm may readily be seen in specific cases.

To illustrate this point, we modeled full-scale fire tests using both the old and the new algorithms. The test involved fc the Navy's fire test ship, ex-USS SHADWELL, with a diesel spray fuel fire in compartment 1. For these tests, pilot f spray fires were ignited at +183 s. The configuration is illustrated in C-2; the double-headed arrows indicate vents b compartment and the exterior. A complete description of the test is given in Bailey and Tatem [A6].

Figure A-3 shows that the new absorption calculation had relatively little effect on the predicted upper layer tempera very sooty fires and, for that compartment, the absorption coefficient value assumed by the original algorithm was a g predicted the observed air temperatures.

For comparison, we have included the results obtained with the new algorithm when soot generation is turned off (by zero). Compared with the case of a realistic amount of soot, CFAST predicts much higher upper layer temperatures qualitatively explained as follows:

The only source of energy in the fire compartment is the fire, modeled as a plume which injects the combustion prod constant for the two cases, the same amount of energy is delivered to the upper layer in both cases. The upper layer conduction of energy. The latter two are unaffected by any change in the absorbance, but the radiative losses will be that, for a gray body, the emissivity is the same as the absorptivity) when soot is eliminated. With smaller radiative los

112

therefore the no-soot gas temperature will be higher than in the baseline case.

The situation in the upper layer of compartment 2 is different from that in the upper layer of the fire compartment. Th from the hot deck so the temperature is directly related to layer absorbance. As seen in Figure A-4, the new algoritl brings it into closer agreement with experimental values. For the no-soot case, absorbance is significantly reduced, re predictions.

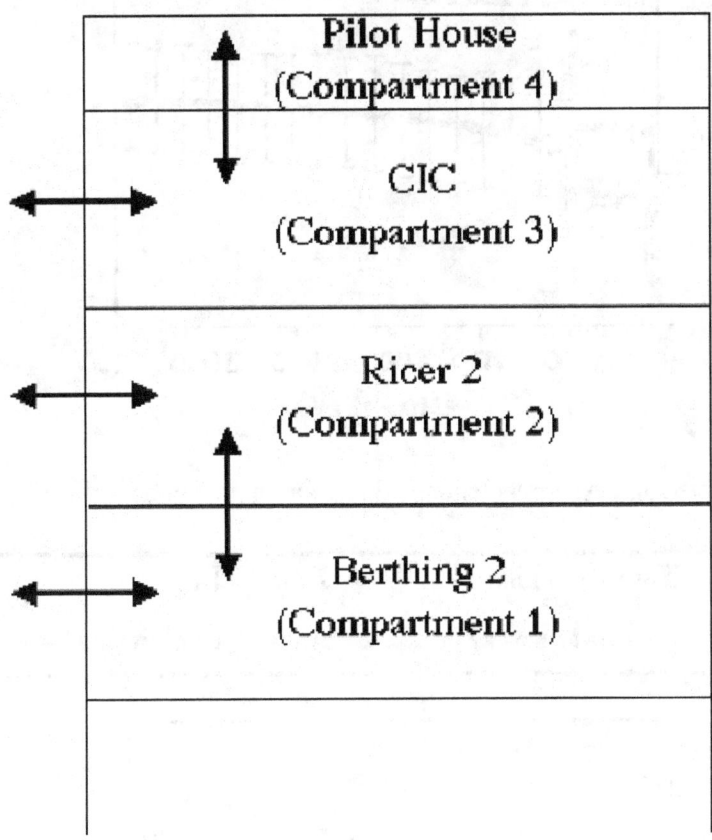

Figure A-2. Schematic of shipboard full-scale fire test area.

In this case, the vertical vent between compartments 1 and 2 was very small (0.002 m²), which had two effects: (a) r energy transport mechanism within compartment 2; and (b) very little soot was present in compartment 2. As a cons overestimated the absorbance of the upper layer and overpredicted the layer temperature. The new algorithm better concentration and comes closer to the actual temperatures. As we would expect, even less absorbance would occur observed temperatures. This interpretation is consistent with the absorption coefficients shown in Figure A-5. The n to the H_2O and CO_2 contributions.

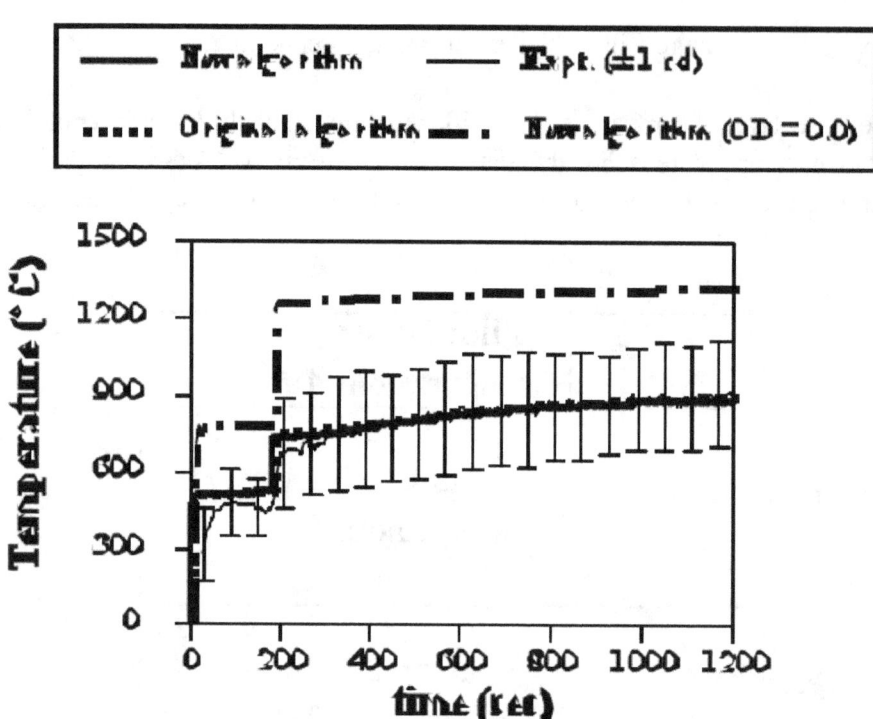

Figure A-3. Predicted and experimental upper layer air temperatures for compartment 1.

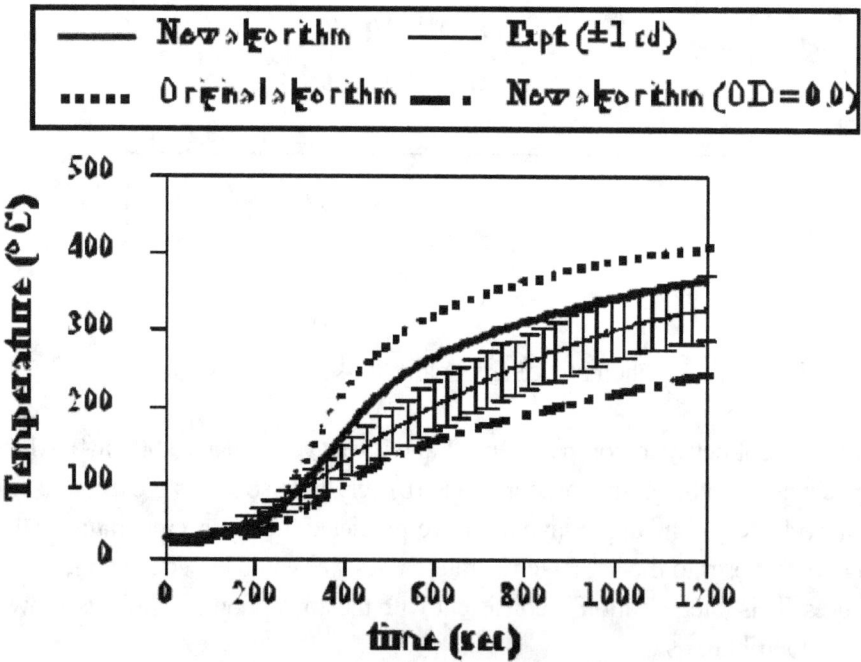

Figure A-4. Predicted and experimental upper layer air temperatures for compartment 2.

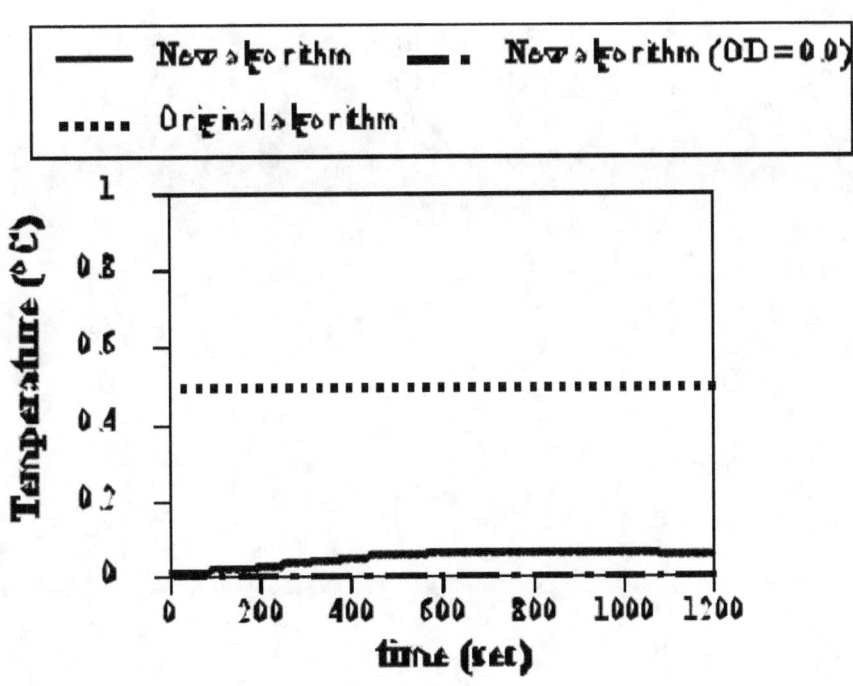

Figure A-5. Predicted upper layer absorption coefficient for compartment 2.

Mathematical symbols

Symbol	Meaning	Units
A	Layer surface area	meter^2
A	Absorbance	dimensionless
α	Specific absorption coefficient	(meter-concentration)$^{-1}$
C	Concentration	unspecified
c	Mean beam length constant	dimensionless
f_V	Soot volume fraction	dimensionless
k	Optical constant for soot	(meter-Kelvin)$^{-1}$
κ	Mean absorption coefficient	meter^{-1}
L	Path length	meter
S	Scattering fraction	dimensionless
T	Temperature	Kelvin
T	Transmittance	dimensionless
V	Layer volume	meter^3

Subscript   Meaning

EFF	Effective
G	Gas
T	Total
S	Soot

References

[A1]   Tien *et. al.*, "Radiation Heat Transfer" in DiNenno, P.J. (Ed.) *The SFPE Handbook of Fire Protection Engineering*, 1st Ed., National Fire Protection Association, Eq. 34, p. 1-99 (1988).

[A2]   Edwards, D.K., "Radiation Properties of Gases" In Rohsenow, W.M., *Handbook of Heat Transfer Fund* Ed., Chap. 14, Figures 2 and 3, pp. 74-75 (1985).

[A3]   Tien, *et. al.*, "Radiation Heat Transfer." in DiNenno, P.J. (Ed.) *The SFPE Handbook of Fire Protection Engineering*, 1st Ed., National Fire Protection Association, Eq. 39, p. 1-100 (1988).

[A4]   Hubbard, GL. and Tien, C.L. "Infrared Mean Absorption Coefficients of Luminous Flames and Smoke," *J.* (1978).

[A5]   Tien, *et. al.*, "Radiation Heat Transfer." in DiNenno, P.J. (Ed.) *The SFPE Handbook of Fire Protection Engineering*, 1st Ed., National Fire Protection Association, Table 1-52, p. 1-98 (1988).

[A6]   Bailey, J.L. and Tatem, P.A., "Validation of Fire/Smoke Spread Model (CFAST) Using Ex-USS SHADWE Control (ISCC) Fire Tests," US Naval Research Lab. Memo Report 6180-95-7781 (1995) [NTIS No. AI

Appendix B CFAST Input Data File Format

The CFAST model requires a description of the problem to be solved. This section provides a description for the input data used by the model. In general, the order of the data is not important. The one exception to this is the first line which specifies the version number and gives the data file a title.

Most entries in the input data file can be generated using FAST. FAST provides online help information in addition to context-sensitive error checking. For example, FAST will not allow the user to select a fire compartment outside the range of compartments specified on the Geometry screen. Because of these features, FAST is the preferred method for creating and editing most CFAST input files. However, some input file key words are not supported by FAST. For these special cases, editing of the ASCII input file using any ASCII text editor is necessary. The following sections detail the available input file key words and group them by their availability within FAST. Subsection titles for the FAST key words correspond to the subsection titles in the FAST chapter of this reference. This has been done as an aid to understanding the organization of the input file.

The number of lines in a given data set will vary depending, for example, on the number of openings or the number of species tracked. A number of parameters such as heat transfer and flow coefficients have been set within CFAST as constants.

B.1 General Format of an Input File Line

Each line of the data file begins with a key word which identifies the type of data on the line. The key words currently available are listed below. The maximum number of arguments for each keyword is shown in parentheses at the right of the description for the keyword.

ADUMPF	specify a file name for saving time histories in spreadsheet form	(2)
CEILI	specify name of ceiling descriptor(s)	(N)
CFCON	ceiling floor heat conduction	(2)
CHEMI	miscellaneous parameters for kinetics	(7)
CJET	ceiling jet	(1)
CO	CO/CO_2 mass ratio	(lfmax)
CT	fraction of fuel which is toxic	(lfmax)
CVENT	opening/closing parameter	(lfmax + 3)
DEPTH	depth of compartments	(N)
DETECT	fire detection and suppression	(9)
DUMPR	specify a file name for saving time histories	(1)
EAMB	external ambient	(3)
FAREA	area of the base of the fire	(lfmax)

FHIGH	height of the base of the fire	(lfmax)
FLOOR	specify the name of floor property descriptor(s)	(N)
FMASS	pyrolysis rate	(lfmax)
FPOS	exact position of the fire using x, y, z coordinates	(3)
FQDOT	heat release rate	(lfmax)
FTIME	points of time on the fire timeline	(Lfmax + 1)
HALL	specify corridor flow model	(1)
HCL	hcl/pyrolysis mass ratio	(lfmax)
HCN	hcn/pyrolysis mass ratio	(lfmax)
HCR	hydrogen/carbon mass ratio of the fuel	(lfmax)
HEIGH	interior height of a compartment	(N)
HHEAT	heat transfer between connected compartment walls	(2 x N)
HI/F	absolute height of the floor of a compartment	(N)
HVENT	specify vent which connect compartments horizontally	(7)
INELV	specify **interior** node elevations (for ventilation ducts)	(2 x # of interior nodes)
LFBO	compartment of fire origin	(1)
LFBT	type of fire	(1)
LFPOS	position of the fire in the compartment	(1)
MVDCT	describe a piece of (circular) duct work	(9)
MVFAN	give the pressure - flow relationship for a fan	(5 to 9)
MVOPN	Specify an opening between a compartment and ventilation system	(5)
OBJECT	additional objects to be burned	(7)
OBJFL	alternative object database file	(1)
OD	C/CO_2 mass ratio	(lfmax)
O2	ratio of oxygen to carbon in the fuel	(lfmax)
RESTR	specify a restart file	(2)
ROOMA	specify room cross-sectional area as a function of height	(lfmax)
ROOMH	specify room heights corresponding to areas specified with ROOMA	(lfmax)
SELECT	specify compartments for graphical display in GUI interface	(3)
SHAFT	specify single zone model for a compartment	(1)
STPMAX	specify maximum ODE solver timestep	(1)
TAMB	ambient inside the structure	(3)
TARG	specify a simplified wall surface target	
TARGET	specify targets for calculation of local surface temperature and flux	(10)
THRMF	alternative thermal properties file	(1)
TIMES	time step control of the output	(5)
VERSN	version number and title	(fixed format 2)
VVENT	specify a vent which connects compartments vertically	(4)
WALLS	specify the name of wall property descriptor(s)	(N)
WIDTH	width of the compartments	(N)
WIND	scaling rule for wind effects	(3)

The number in parenthesis is the maximum number of entries for that line. "N" represents the number of compartments being modeled. The outside (ambient) is designated by one more than the number of compartments, N+1. Thus, a three compartment model would refer to the outside as compartment four. An entry for lfmax is no longer supported directly. The value for lfmax is determined by the number of entries on the FTIME line.

Each line of input consists of a label followed by one or more alphanumeric parameters associated with that input label. The label must always begin in the first space of the line and be in capital letters. Following the label, the values may start in any column, and all values must be separated by either a comma or a space. Values may contain decimal points if needed or desired. They are not required. Units are standard SI units. Most parameters have default values which can be utilized by omitting the appropriate line. These are indicated in the discussion. The maximum line length is 128 characters, so all data for each key word must fit in this number of characters.

B.2 Entering a Title for the Input File

The first line in the file must be the version identification along with an optional short description for the simulation. It is a required input. The VERSN line is the line that CFAST keys on to determine whether it has a correct data file. The format is fixed, that is the data must appear in the columns specified in the text.

Example:

```
VERSN    1 Example Case for FAST User's Guide
```

Key word: VERSN	
Inputs: Version Number, Title	
Version Number	The version number parameter specifies the version of the CFAST model for which the input data file was prepared. Normally, this would be 3. It must be in columns 8-9.
Title	The title is optional and may consist of letters, numbers, and/or symbols that start in column 11 and may be up to 50 characters. It permits the user to uniquely label each run.

B.3 Specifying Simulation and Output Times

A TIMES line is also required in order to specify the length of time over which the simulation takes place.

Example:

```
VERSN    1 Example Case for CFAST 1.6 User's Guide
TIMES    200    10    10    0    0
```

Key word: TIMES	
Inputs: Simulation time, Print Interval, History Interval, Display Interval, Copy Count	
Simulation Time (s)	Simulation time is the length of time over which the simulation takes place. The maximum value for this input is 86400 s (1 day). The simulation time parameter is required.
Print Interval (s)	The print interval is the time interval between each printing of the output values. If omitted or less than or equal to zero, no printing of the output values will occur.
History Interval (s)	The history interval is the time interval between each writing of the output to the history file. The history file stores all of the output of the model at the specified interval in a format which can be efficiently retrieved for use by other programs. Section B.13 provides details of the history file. A zero must be used if no history file is to be used.
Display Interval (s)	The display interval is the time interval between each graphical display of the output as specified in the graphics specification, section A.16. If omitted, no graphical display will occur. There is a maximum of 900 intervals allowed. If the choice for this parameter would yield more than 900 writes, the graphs are truncated to the first 900 points.
Copy Count	Copy count is the number of copies of each graphical display to be made on the selected hard copy device as specified in the graphics specification, section A.16. If omitted, a value of zero (no copies) is assumed.

When running simulations in the GUI interface, up to three compartments may be selected and displayed in tabular and graphical form as the simulation proceeds. The SELECT keyword specifies the user's choice for the (up to) three compartments.

Key word: SELECT	
Input: Compartment 1, Compartment 2, Compartment 3	
Compartment 1 Compartment 2 Compartment 3	Each of up to three inputs may be included on the input line to specify the compartments to be displayed in tabular, graphical, and spreadsheet form during the simulation. The compartment numbers may be in any order and any compartments included in the simulation may be selected. Entries less than one, or greater than the number of compartments in the simulation are ignored.

Example:

```
SELECT 3 1 2
SELECT 1 0 0
```

There are several files which CFAST uses to communicate with its environment. They include 1) a configuration file, 2) the thermal database, 3) the objects database, 4) a history file, and 5) a restart file. The format of the thermal database and objects database are detailed in Appendices B and C.

The output of the simulation may be written to a disk file for further processing by programs such as REPORT or to restart the CFAST model from the end of a previous simulation. At each interval of time as specified by the history interval in the TIMES label, the output is written to the file specified. For efficient disk storage and optimum speed, the data is stored in an internal format and cannot be read directly with a text editor. The RESTR line is an optional line used to restart the model at a specified simulation time within an existing history file.

Example:

```
DUMPR PRM.HI
ADUMPF PRM.CSV NF
```

Key word: ADUMPF Input: Spreadsheet File, Output Options	
Spreadsheet File	The name specifies a file (up to 63 characters) to which the program outputs are written in a form readable by spreadsheet programs. Spreadsheet file is an optional input. If omitted, the file will not be generated. Note that in order to obtain a spreadsheet file, this parameter must be specified, and the display interval (see Section B.2) must be set to a non-zero value.
Output Options	Specifies the type of output(s) written to the spreadsheet file. One or more of the letters WINFS may be specified. See Appendix B.

Key word: DUMPR Input: History File	
History File	The name specifies a file (up to 17 characters) to which the program outputs for plotting are written. History file is an optional input. If omitted, the file will not be generated. Note that in order to obtain a history of the variables, this parameter must be specified, and the history interval (see Section B.2) must be set to a non-zero value.

Key word: RESTR Input: Restart File, Restart Time (see Section B.2)	
Restart File	The name specifies a file (up to 17 characters) from which the program reads data to restart the model. This data must have been generated (written) previously with the history parameter discussed earlier.
Restart Time (s)	A time step is given after the name of the file and specifies at what time the restart should occur.

B.4 Setting Ambient Conditions

The ambient conditions section of the input data allows the user to specify the temperature, pressure, and station elevation of the ambient atmosphere, as well as the absolute wind pressure to which the

123

structure is subjected. There is an ambient for the interior and for the exterior of the structure. The key word for the interior of the structure is TAMB and for the exterior of the structure is EAMB. The form is the same for both.

The key word for the wind information is WIND. The wind modification is applied only to the vents which lead to the exterior. Pressure interior to a structure is calculated simply as a lapse rate based on the NOAA tables. This modification is applied to the vents which lead to the exterior ambient. The calculated pressure change is modified by the wind coefficient for each vent. This coefficient, which can vary from -1.0 to +1.0, nominally from -0.8 to +0.8, determines whether the vent is facing away from or into the wind. The pressure change is multiplied by the vent wind coefficient and added to the external ambient for each vent which is connected to the outside.

The choice for the station elevation, temperature and pressure must be consistent. Outside of that limitation, the choice is arbitrary. It is often convenient to choose the base of a structure to be at zero height and then reference the height of the structure with respect to that height. The temperature and pressure must then be measured at that position. Another possible choice would be the pressure and temperature at sea level, with the structure elevations then given with respect to mean sea level. This is also acceptable, but somewhat more tedious in specifying the construction of a structure. Either of these choices works though because consistent data for temperature and pressure are available from the Weather Service for either case.

If the EAMB or TAMB line is not included in the input file, default values are used. The WIND line is optional.

Example:

```
TAMB   300.   101300.      0.
EAMB   300.   101300.      0.
```

Key words: EAMB and TAMB	
Inputs: Ambient Temperature, Ambient Pressure, Station Elevation	
(External and Internal, respectively)	
Ambient Temperature (K)	Ambient temperature is the temperature of the ambient atmosphere. Default is 300.
Ambient Pressure (Pa)	The ambient pressure is the pressure of the ambient atmosphere. Default is 101300.
Station Elevation (m)	The station elevation is the elevation of the point at which the ambient pressure and temperature (see above) are measured. The reference point for the elevation, pressure and temperature must be consistent. This is the reference datum for calculating the density of the atmosphere as well as the temperature and pressure inside and outside of the structure as a function of height. Default is 0.

Key word: WIND	
Inputs: Wind Speed, Reference Height, Lapse Rate Coefficient	
Wind Speed (m/s)	Wind speed at the reference elevation. The default is 0.
Reference Height (m)	Height at which the reference wind speed is measured. The default is 10 m.
Lapse Rate Coefficient	The power law used to calculate the wind speed as a function of height. The default is 0.16.

B.5 Defining Compartments

This section allows the user to portray the geometry of the structure being modeled. The size and location of every compartment in the structure MUST be described. The maximum number of compartments is 15 compartments (plus the outdoors). The structure of the data is such that the compartments are described as entities, and then connected in appropriate ways. It is thus possible to have a set of compartments which can be configured in a variety of ways. In order to specify the

125

geometry of a structure, it is necessary to give the physical characteristics. Thus the lines labelled HI/F, WIDTH, DEPTH and HEIGH are all required. Each of these lines requires "N" data entries, that is one for each compartment.

Example:

```
WIDTH    4.00    4.00
DEPTH    4.00    4.00
HEIGH    2.30    2.30
HI/F     0.00    2.30
```

Key word: WIDTH Input: Compartment Width	
Compartment Width (m)	Compartment width specifies the width of the compartment. The number of values on the line must equal the number of compartments in the simulation.

Key word: DEPTH Input: Compartment Depth	
Compartment Depth (m)	Compartment depth specifies the depth of the compartment. The number of values on the line must equal the number of compartments in the simulation.

Key word: HEIGH Input: Compartment Height	
Compartment Height (m)	Compartment Height specifies the height of the compartment. The number of values on the line must equal the number of compartments in the simulation.

Key word:	HI/F
Input: Floor Height	
Floor Height (m)	The floor height is the height of the floor of each compartment with respect to station elevation specified by the TAMB parameter. The reference point must be the same for all elevations in the input data. The number of values on the line must equal the number of compartments in the simulation.

Two additional commands, ROOMA and ROOMH, may be used for defining compartment properties. The ROOMA and ROOMH commands allow the user to define non-rectangular rooms by specifying cross-sectional area as a function of height.

Example:

```
ROOMA 1    3      10.0  5.0  3.0
ROOMH 1    3      0.0   1.0  2.0
```

The above example specifies that compartment 1 has a cross-sectional area of 10, 5 and 3 m^2 at elevations 0.0, 1.0 and 2.0 m respectively.

Key word: ROOMA	
Input: First Compartment, Number of Area Data Points, Area Data Point(s)	
First Compartment	The compartment in which the cross-sectional area varies as a function of height
Number of Area Data Points	A varying cross-sectional area is specified by including a specified number of cross-sectional area values as a function of height, with matching height values specified with the ROOMH command. The number of area data points specified must equal the number of matching cross-sectional area values specified below. The number of values for the compartment must match those included for the same compartment on the ROOMH command, below
Area Data Point(s)	Values of cross-sectional area of the compartment as a function of height measured from the floor of the compartment. The values for the compartment correspond to height values included for the same compartment on the ROOMH command, below

Key word: ROOMH	
Input: First Compartment, Number of Height Data Points, Height Data Point(s)	
First Compartment	The compartment in which the cross-sectional area varies as a function of height
Number of Height Data Points	A varying cross-sectional area is specified by including a specified number of height values with matching cross-sectional area values specified with the ROOMA command. The number of height data points specified must equal the number of matching cross-sectional area values specified above. The number of values for the compartment must match those included for the same compartment on the ROOMA command, above
Height Data Point(s)	Values of height for the corresponding cross-sectional area values measured from the floor of the compartment. The values for the compartment coorespond to cross-sectional area values included for the same compartment on the ROOMA command, above.

B.6 Thermal Properties

The thermophysical properties of the enclosing surfaces are described by specifying the thermal conductivity, specific heat, emissivity, density, and thickness of the enclosing surfaces for each compartment. Currently, thermal properties for materials are read from a thermal database file unique to CFAST. The data in the file simply gives a name (such as CONCRETE) which is a pointer to the properties in the thermal database. The thermophysical properties are specified at *one* condition of temperature, humidity, *etc*. There can be as many as three layers per boundary, but they are specified in the thermal database itself.

If the thermophysical properties of the enclosing surfaces are not included, CFAST will treat them as adiabatic (no heat transfer). If a name is used which is not in the database, FAST will turn off the conduction calculation, and CFAST will stop with an appropriate error message.

Since most of the heat conduction is through the ceiling, and since the conduction calculation takes a significant fraction of the computation time, it is recommended that initial calculations be made using the ceiling only. Adding the walls generally has a small effect on the results, and the floor contribution is usually negligible. Clearly, there are cases where the above generalization does not hold, but it may prove to be a useful screening technique.

The default name for the thermal properties database is THERMAL.TPF. Another name can be used by selecting it during installation, or by using the key word THRMF in the CFAST data file.

Example:

```
CEILI GYPSUM    GYPSUM
WALLS PINEWOOD  PINEWOOD
FLOOR CONCRETE  CONCRETE
```

Key word: CEILI Inputs: Ceiling Materials	
Ceiling Materials	The label CEILI indicates that the names of thermophysical properties on this line describe the ceiling material. If this parameter is present, there must be an entry for each compartment.

Key word: WALLS Inputs: Wall Materials	
Wall Materials	The label WALLS indicates that the names of thermophysical properties on this line describe the wall material. If this parameter is present, there must be an entry for each compartment.

Key word: FLOOR Inputs: Floor Materials	
Floor Materials	The label FLOOR indicates that the names of thermophysical properties on this line describe the floor material. If this parameter is present, there must be an entry for each compartment.

Key word: THRMF Input: Thermal Database	
Thermal Database	The name specifies a file (up to 12 characters) from which the program reads thermophysical data. If this parameter is not specified, then either the default (THERMAL.DAT) is used, or the name is read from the configuration file.

B.7 Defining Connections for Horizontal Flow

This section of the input data file is required to specify horizontal flow connections between compartments in the structure. These may include doors between compartments or windows in the compartments (between compartments or to the outdoors). These specifications do **not** correspond to physically connecting the walls between specified compartments. Lack of an opening prevents flow between the compartments. Openings to the outside are included as openings to a compartment with a number one greater than the number of compartments described in the Geometry section. The key word is HVENT. If the HVENT line is entered, the first six entries on the line are required. There is an optional seventh parameter to specify a wind coefficient. The soffit and sill specifications are with respect to the first compartment specified and is not symmetric. Reversing the order of the compartment designations does make a difference.

Horizontal flow vents may be opened or closed during the fire with the use of the CVENT key word. The initial format of CVENT is similar to HVENT specifying the connecting compartments and vent number. Each CVENT line in the input file details the open/close time dependent characteristics for one horizontal flow vent by specifying a fractional value for each LFMAX time entry. The default is 1.0 which is a fully open vent. A value of 0.5 would specify a vent which was halfway open.

Example:

```
HVENT   1  3  1    1.07    2.00    0.00    0.00
HVENT   2  3  1    1.07    2.00    1.00    0.00
CVENT   1  3  1    1.00    1.00
CVENT   2  3  1    1.00    1.00
```

Key word: HVENT	
Inputs: First Compartment, Second Compartment, Vent Number, Width, Soffit, Sill, Wind	
First Compartment	The first compartment is simply the first connection.
Second Compartment	The second compartment is the compartment number to which the first compartment is connected. The order has one significance. The height of the sill and soffit are with respect to the first compartment specified.
Vent Number	There can be as many as four vents between any two compartments. This number specifies which vent is being described. It can range from one to four.
Width (m)	The width of the opening.
Soffit (m)	Position of the top of the opening above the floor of the compartment number specified as the first compartment.
Sill (m)	Sill height is the height of the bottom of the opening above the floor of the compartment number specified as the first compartment.
Wind	Optional parameter, the wind coefficient is the cosine of the angle between the wind vector and the vent opening. This applies only to vents which connect to the outside ambient (specified with EAMB). The range of values is -1.0 to +1.0. If omitted, the value defaults to zero.
First Compartment Position	Optional parameter, horizontal distance between the centerline of this vent and the reference point in the first compartment
Second Compartment Position	Optional parameter, horizontal distance between the centerline of this vent and the reference point in the second compartment

Key word: CVENT	
Inputs: First Compartment, Second Compartment, Vent Number, Width	
First Compartment	The first compartment.
Second Compartment	The second compartment is the compartment number to which the first compartment is connected.
Vent Number	This number specifies which vent is being described. It can range from one to four.
Width	Fraction that the vent is open. This applies to the width only. The sill and soffit are not changed. The number of values on the line must equal the number of points on the fire timeline.

B.8 Defining Connections for Vertical Flow

The Vents(ceiling,...) section of the input data file describes any vertical flow openings, such as scuddles, between compartments in the structure (or between a compartment and the outdoors). Openings to the outside are included as openings to a compartment with a number one greater than the number of compartments described in the Geometry section. Each VVENT line in the input file describes one vertical vent. There are four parameters, the connected compartments, the shape of the opening, and the effective area of the vent. At the present time, there is not an equivalent CVENT mechanism for opening or closing the vertical vents.

Example:

```
VVENT   2   1  1.00   1
```

Key word: VVENT	
Inputs: First Compartment, Second Compartment, Area, Shape	
First Compartment	The first compartment is simply the first connection.
Second Compartment	The second compartment is the compartment number to which the first compartment is connected.
Area (m^2)	This is the effective area of the opening. For a hole, it would be the actual opening. For a diffuser, the effective area is somewhat less than the geometrical size of the opening.

Shape	1 for circle or 2 for square.

B.9 Adding Sprinklers and Detectors

Sprinklers and detectors are both considered detection devices by the FAST model and are handled using the same input keywords. Detection is based upon heat transfer to the detector. Fire suppression by a user-specified water spray begins once the associated detection device is activated. A maximum of 20 sprinklers or detectors can be included for any input file and model run. These can be in one compartment or scattered throughout the structure. The DETECT keyword is used for both detectors and sprinklers.

Key word: DETECT Inputs: Detector Type, Compartment, Activation Temperature, Depth Position, Breadth Position, Height Position, RTI, Sprinkler, Spray Density	
Detector Type	Type of detector, 1 for smoke detector and 2 for heat detector.
Compartment	The compartment in which the detector or sprinkler is located.
Activation Temperature (K)	The temperature at or above which the detector link activates.
Depth Position (m)	Position of the detector or sprinkler as a distance from the rear wall of the compartment (X direction). Default value is ½ compartment depth.
Breadth Position (m)	Position of the object as a distance from the left wall of the compartment (Y direction). Default value is ½ compartment breadth.
Height Position (m)	Position of the object as a distance from the floor of the compartment (Z direction). Default value is compartment depth.
RTI ($(m \cdot s)^{1/2}$)	The Response Time Index (RTI) for the sprinkler or detection device.
Sprinkler	If set to a value of 1, the sprinkler will quench the fire with the specified spray density of water.

134

Spray Density (m/s)	The amount of water dispersed by a water spray-sprinkler. The units for spray density are length/time. These units are derived by dividing the volumetric rate of water flow by the area protected by the water spray. The spray density may be measured by collecting water in a pan located within the spray area and recording the rate-of-rise in the water level.

B.10 Defining the Fire and Time-Dependent Fire Curves

The fire specifications allow the user to describe the fire source in the simulation. The location and position of the fire are specified using the LFBO and FPOS lines. Chemical properties of the fuel are specified with the CHEMI key word along with miscellaneous parameters. Turn the ceiling jet calculations on by using the CJET key word. By default, the ceiling jet is not included in a CFAST simulation.

By default, the fire is placed in the center of the compartment on the floor. To place the fire in a different location, the FPOS key word may be included in the input file. If values for any of the three variables are invalid (i.e., less than zero or greater than the compartment dimension in the appropriate direction), the location for that direction defaults to the center of the appropriate direction.

CFAST no longer supports use of the LFMAX key word in the data file. LFMAX is now determined by the number of entries on the FTIME line used to specify points of the fire timeline. The time dependent variables of the fire are described with a series of mass loss rate, rate of heat release, fuel height, and fuel area inputs. All of these specifications are optional. If entered, a total of LFMAX+1 values must be included for each time dependent input line. The defaults shown for each key word reflect the values for methane.

With the three parameters, the heat of combustion (HOC) from CHEMI, FMASS and FQDOT, the pyrolysis and heat release rate are over specified. The model uses the last two of the three to obtain the third parameter. That is, if the three were specified in the order HOC, FMASS and FQDOT, then FQDOT would be divided by FMASS to obtain the HOC for each time interval. If the order were FMASS, FQDOT and HOC, then the pyrolysis rate would be determined by dividing the heat release rate by the heat of combustion. If only two of the three are given, then those two will determine the third, and finally, if none or only one of the parameters is present, the defaults shown are used.

Species production rates are specified in a manner similar to the fire, entering the rates as a series of points with respect to time. The species which are followed by CFAST are:

- Carbon Dioxide
- Carbon Monoxide

- Concentration-Time Product
- Hydrogen Cyanide
- Hydrogen Chloride
- Nitrogen
- Oxygen
- Soot (Smoke Density)
- Total Unburned Hydrocarbons
- Water

The program performs a linear interpolation between the time points to determine the time of interest.

For a type one (LFBT=1) fire, only the concentration-time product of pyrolysate (CT) can be specified. No other species are followed. For a type two (LFBT=2) fire, nitrogen (N2), oxygen (O2), carbon dioxide (CO2), carbon monoxide (CO), hydrogen cyanide (HCN), hydrogen chloride (HCL), soot (OD), unburned fuel and water (H2O) are followed. For a type two fire, HCN, HCL, CT, O2, OD, CO and the hydrogen to carbon ratio (HCR) can be specified. In all cases, the unit of the production rates is kg/kg. However, the meaning of the production rates is different for the several types of species. See the discussion for each species key word below for the meaning of the corresponding production rate.

Example:

```
CHEMI  16.     0.  10.0     18100000.  300. 400.   0.
LFBO   1
LFBT   2
FPOS   2.00    2.00    0.00
FTIME  400.
FMASS  0.0014 0.0014
FHIGH  0.00    0.01
FAREA  0.00    0.00
FQDOT  2.53E+04 2.53E+04
CJET   OFF
HCR    0.333 0.333
CO     0.010 0.010
```

Key word: CHEMI	
Inputs: Molar Weight, Relative Humidity, Lower Oxygen Limit, Heat of Combustion, Initial Fuel Temperature, Gaseous Ignition Temperature, Radiative Fraction	
Molar Weight	Molecular weight of the fuel vapor. This is the conversion factor from mass density to molecular density for "tuhc." Default is 16. It is used only for conversion to ppm, and has no effect on the model itself.
Relative Humidity (%)	The initial relative humidity in the system. This is converted to kilograms of water per cubic meter.
Lower Oxygen Limit (%)	The limit on the ratio of oxygen to other gases in the system below which a flame will not burn. This is applicable only to type (LFBT) 2 or later fires. The default is 10.
Heat of Combustion (J/kg)	Heat of combustion of the fuel. Default is 50000000.
Initial Fuel Temperature (K)	Typically, the initial fuel temperature is the same as the ambient temperature as specified in the ambient conditions section.
Gaseous Ignition Temperature (K)	Minimum temperature for ignition of the fuel as it flows from a compartment through a vent into another compartment. If omitted, the default is arbitrarily set to the initial fuel temperature plus 100K.
Radiative Fraction	The fraction of heat released by the fire that goes into radiation. Default is 0.30.

Key word: FAREA	
Inputs: Fuel Area	
Fuel Area (m)	The area of the fire at the base of the flames.

Key word: FHIGH	
Inputs: Fuel Height	

Fuel Height (m)	The height of the base of the flames above the floor of the compartment of fire origin for each point of the specified fire.

Key word: FMASS Inputs: Mass Loss Rate	
Mass Loss Rate (kg/s)	The rate at which fuel is pyrolyzed at times corresponding to each point of the specified fire.

Key word: FPOS Inputs: Depth, Breadth, Height (relative to the left rear corner of the compartment – see figure above)	
Depth	Position of the fire as a distance from the rear wall of the compartment (X direction).
Breadth	Position of the fire as a distance from the left wall of the compartment (Y direction).
Height	Height of the fire above the floor (Z direction). This value is simply added to the fire height at each time specified by the FHIGH key word.

Key word: FQDOT Inputs: Heat Release Rate	
Heat Release Rate (W)	The heat release rate of the specified fire.

Key word:	FTIME
Inputs:	Time Points

Time Points (s)	An entry indicates a point on the timeline where mass loss rate, fuel height and species are specified for the fire. This time is independent of the simulation time which is specified for the TIMES label. If the simulation time is longer than the total duration of the fire, the final values specified for the fire (mass loss rate, fuel height, fuel area, and species) are continued until the end of the simulation.

Key word:	LFBO
Input:	Compartment of Fire Origin

Compartment of Fire Origin	Compartment of fire origin is the compartment number in which the fire originates. Default is 0. The outside can not be specified as a compartment. An entry of 0 turns off the main fire leaving only object fires specified by the OBJECT key word.

Key word:	LFBT
Input:	Fire Type

Fire Type	This is a number indicating the type of fire. 1 Unconstrained fire 2 Constrained fire. The default is 1. See section 3.10 for a discussion of the implications of this choice.

Key words:	HCN, HCL, CT, HCR, or O2
Inputs:	Composition of the Pyrolyzed Fuel

Production Rate (kg/kg)	Units are kilogram of species produced per kilogram of fuel pyrolyzed for HCN and HCL. The input for CT is the kilograms of "toxic" combustion products produced per kilogram of fuel pyrolyzed. Input for HCR is the mass ratio of hydrogen to carbon ratio in the fuel. Input for O2 is the mass ratio of oxygen to carbon *as it becomes available from the fuel*. The O2 input thus represents excess oxygen available from the fuel for combustion. For normal fuels, this input should be left out.

Key words: OD and CO Inputs: Yield	
Yield (kg/kg)	Input the ratio of the mass of carbon to carbon dioxide produced by the oxidation of the fuel for OD. The input for CO is the ratio of the mass of carbon monoxide to carbon dioxide produced by the oxidation of the fuel.

B.11 Building HVAC Systems

These key words are used to describe a mechanical ventilation system. The MVOPN line is used to connect a compartment to a node in the mechanical ventilation system. The elevation for each of these exterior nodes is specified as a relative height to the compartment floor on the MVOPN line. The MVDCT key word is used to specify a piece of the mechanical ventilation duct work. CAUTION: Nodes specified by each MVDCT entry must connect with other nodes, fans, or compartments. Do not specify ducts which are isolated from the rest of the system. Specify interior elevations of the mechanical ventilation nodes using the INELV line. All node elevations can be specified, but elevations for the exterior nodes, that is those connected to a compartment, are ignored. These heights are determined by entries on the MVOPN line. The heights for interior nodes are absolute heights above the reference datum specified by TAMB. The heights are specified in pairs with the node number followed by the height.

A fan is defined using the MVFAN line to indicate node numbers and to specify the fan curve with power law coefficients. There must be at least one and a maximum of five coefficients specified for each MVFAN entry. The fan coefficients are simply the coefficients of an interpolating polynomial for the flow speed as a function of the pressure across the fan housing. In this example, the coefficients:

```
B( 1) =    0.140E+00      b(1)
B( 2) =   -0.433E-03      b(2) x p
```

140

were calculated from entries made in FAST:

```
            PRESSURE        FLOW

Minimum       0.00         0.1400
Maximum     300.00         0.0101
```

Example:

```
MVOPN   1   1 V   2.10    0.12
MVOPN   2   3 V   2.10    0.12
MVDCT   1   2   2.30   0.10 .00200   0.00 1.0000    0.00 1.0000
MVFAN   2   3   0.00 300.00   0.140E+00 -0.433E-03
INELV   1   2.10    2   4.40    3   4.40
```

Key word: INELV Inputs: Node Number, Height	
Node Number	Number of an interior node.
Height (m)	Height of the node with respect to the height of the reference datum, specified by TAMB or EAMB.

Key word: MVDCT	
Inputs: First Node Number, Second Node Number, Length, Diameter, Absolute Roughness, First Flow Coefficient, First Area, Second Flow Coefficient, Second Area	
First Node Number	First node number. This is a node in the mechanical ventilation scheme, not a compartment number (see MVOPN).
Second Node Number	Second node number.
Length (m)	Length of the duct.
Diameter (m)	All duct work is assumed to be circular. Other shapes must be approximated by changing the flow coefficient. This is done implicitly by network models of mechanical ventilation and forced flow, but must be done explicitly here.
Absolute Roughness (m)	Roughness of the duct.
First Flow Coefficient	Flow coefficient to allow for an expansion or contraction at the end of the duct which is connected to node number one. To use a straight through connection (no expansion or contraction) set to zero.
First Area (m²)	Area of the expanded joint.
Second Flow Coefficient	Coefficient for second node.
Second Area (m²)	Area at the second node.

Key word:	MVFAN
Inputs: First Node, Second Node, Minimum Pressure, Maximum Pressure, Coefficients	
First Node	First node in the mechanical ventilation system to which the fan is connected.
Second Node	Second node to which the fan is connected.
Minimum Pressure (Pa)	Lowest pressure of the fan curve. Below this value, the flow is assumed to be constant.
Maximum Pressure (Pa)	Highest pressure at which the fan will operate. Above this point, the flow is assumed to stop.
Coefficients	At least one, and a maximum of five coefficients, to specify the flow as a function of pressure.

Key word:	MVOPN
Inputs: Compartment Number, Duct Work Node Number, Orientation, Height, Area	
Compartment Number	Specify the compartment number.
Duct Work Node Number	Corresponding node in the mechanical ventilation system to which the compartment is to be connected.
Orientation	V for vertical or H for horizontal.
Height (m)	Height of the duct opening above the floor of the compartment.
Area (m^2)	Area of the opening into the compartment.

B.12 Selection of Ceiling Jet Surfaces

Key word: CJET Input: OFF, CEILING, WALL, or ALL	
Current Setting	To include the calculation for the ceiling, wall, or both surfaces, the CJET key word is used together with one of the identifiers CEILING, WALL, or ALL. For example, to turn the ceiling on, use "CJET CEILING". At present, this key word effects only the calculation of the convective heating boundary condition for the conduction routines. If a particular surface is ON, the ceiling jet algorithm is used to determine the convective heating of the surface. If OFF, the bulk temperature of the upper layer determines the convective heating.

B.13 Additional Burning Objects

The OBJECT key word allows the specification of additional objects to be burned in the fire scenario. The object name and object compartment are required if the OBJECT key word is used. All other input items have default values if they are not specified. These defaults are:
start time 0.0, first element 1, depth (x position) one half the depth of the compartment, breadth (y position) one half the width of the compartment, and height (z position) 0.0. To specify any input item, all preceding items on the OBJECT line must also be specified. For example, the first element can not be set if start time is not set. Positioning of the object within a compartment is specified in the same manner as for the main fire. See figure below.

EXAMPLE:

```
OBJECT SOFA       1    10   1   4.00   2.00   0.00
OBJECT WARDROBE   1    30   3   0.00   2.00   0.00
```

Key word: OBJFL Input: Objects Database	
Objects Database	The name specifies a file (up to 17 characters) from which the program obtains object data. If this parameter is not specified, then either the default (OBJECTS.DAT) is used, or the name is read from the configuration file.

Key word:	OBJECT

Inputs: Object Name, Object Compartment, Ignition Criterion Value, Ignition Criterion Type, Depth Position, Breadth Position, Height Position, Normal Vector (Depth), Normal Vector (Breadth), Normal Vector (Height), Horizontal Flame Spread Ignition Position, Vertical Flame Spread Ignition Position

Object Name	The name from the objects database for the desired object. Specifying a name not found in the database causes CFAST to stop with an appropriate error message. FAST considers such an object undefined and does not display the entry.
Object Compartment	The compartment that the object is in during the simulation. If a compartment number outside the range of specified compartments is used, CFAST provides an error message and stops. FAST considers such an object undefined and does not display the entry.
Ignition Criterion Value	The numerical value at which ignition will occur. If it is less than or equal to zero, the default is taken. For constrained and unconstrained fires, the default is and ignition at time of zero. For flame spread objects, the default is the ignition temperature in the database.
Ignition Criterion Type	The type of ignition condition specified by the *Ignition Criterion Value*. Acceptable values are 1 for time, 2 for object surface temperature, and 3 for incident flux to object surface.
Depth Position	Position of the object as a distance from the rear wall of the object compartment (X direction). Default value is ½ compartment depth.
Breadth Position	Position of the object as a distance from the left wall of the object compartment (Y direction). Default value is ½ compartment breadth.
Height Position	Height of the object above the floor (Z direction). Default value is 0.
Normal Vector (Depth)	Specifies a vector of unit length perpendicular to the exposed surface of the object. (Depth) component is in the direction from the rear wall of the object compartment. Default value is a horizontal, upward facing object, unit vector = (0,0,1)
Normal Vector (Breadth)	Specifies a vector of unit length perpendicular to the exposed surface of the object. (Breadth) component is in the direction from the left wall of the object compartment. Default value is a horizontal, upward facing object, unit vector = (0,0,1)

Normal Vector (Height)	Specifies a vector of unit length perpendicular to the exposed surface of the object. (Breadth) component is in the direction from the floor of the object compartment. Default value is a horizontal, upward facing object, unit vector = (0,0,1)
Horizontal Flame Spread Ignition Position	Assuming the ignition surface is a wall with the corner closest to the lower back left corner of the compartment, the horizontal flame spread ignition position is the horizontal distance from the rear lower corner of the object. Only applicable to flame spread objects.
Vertical Flame Spread Ignition Position	Assuming the ignition surface is a wall with the corner closest to the lower back left corner of the compartment, the horizontal flame spread ignition position is the vertical distance from the rear lower corner of the object. Only applicable to flame spread objects.

B.14 Inter-compartment Heat Transfer

Heat transfer between the ceiling and floor of specified compartments is included with the CFCON keyword. Ceiling to floor heat transfer occurs between interior compartments of the structure or between an interior compartment and the outdoors.

Example:

```
CFCON     1     2
```

Key word: CFCON	
Input: First Compartment, Second Compartment	
First Compartment	First of the connected compartments. Order of the inputs is not important
Second Compartment	Second of the connected compartments. Order of the inputs is not important

Horizontal conduction between specified compartments is included with the HHEAT keyword. Ceiling to floor heat transfer occurs between interior compartments of the structure or between an interior compartment and the outdoors.

Examples:

```
HHEAT
HHEAT 2
HHEAT 2    3    1 0.5    3 0.25    4 0.25
```

Key word: HHEAT	
Input: First Compartment, Number of Partrs, Second Compartment, Fraction	
First Compartment	First of the connected compartments. This is an optional parameter. If no arguments are included on the HHEAT line, CFAST computes heat conduction between every pair of rooms connected by a horizontal flow vent. If First Compartment is specified, CFAST calculates heat conduction between that compartment and all compartments connected to First Compartment by a horizontal flow vent.
Number of Parts	Optionally, the user may completely specify the fraction of wall surface connecting the First Compartment and any number of other compartments. The number of partrs specifies the number "second compartment" and "fraction" pairs to be specified below.
Second Compartment	If the number of ordered pairs is specified, pairs of numbers which specify a connected compartment and the fraction of the vertical surface areas of the compartments which are connected can be specified. The second compartment specifies the compartment number of a compartment connected by a wall surface to the first compartment. One pair of numbers should be included for each parts.
Fraction	If the number of ordered pairs is specified, pairs of numbers which specify a connected compartment and the fraction of the vertical surface areas of the compartments which are connected can be specified. The fraction specifies the fraction of the vertical surface area connecting the first and second compartment pair.

B.15 Defining Targets

CFAST can track and report calculations of the heat flux striking and the temperature of arbitrarily positioned and oriented targets. Two keywords are used to specify targets: The TARGET keyword is used to specify arbitrary targets placed anywhere in a compartment. The TARG keyword is used to

147

specify targets on the interior bounding surface of a compartment.

Key word: TARGET Input: Compartment, Depth Position, Breadth Position, Height Position, Normal Vector (Depth), Normal Vector (Breadth), Normal Vector (Height), Material, Method, Equation Type	
Compartment	The compartment in which the target is located
Depth Position	Position of the target as a distance from the rear wall of the target compartment (X direction). Default value is ½ compartment depth.
Breadth Position	Position of the target as a distance from the left wall of the target compartment (Y direction). Default value is ½ compartment breadth.
Height Position	Height of the target above the floor (Z direction). Default value is 0.
Normal Vector (Depth)	Specifies a vector of unit length perpendicular to the exposed surface of the target. (Depth) component in the direction from the rear wall of the target compartment. Default value is a horizontal, upward facing target, unit vector = (0,0,1)
Normal Vector (Breadth)	Specifies a vector of unit length perpendicular to the exposed surface of the target. (Breadth) component in the direction from the left wall of the target compartment. Default value is a horizontal, upward facing target, unit vector = (0,0,1)
Normal Vector (Height)	Specifies a vector of unit length perpendicular to the exposed surface of the target. (Breadth) component in the direction from the floor of the target compartment. Default value is a horizontal, upward facing target, unit vector = (0,0,1)
Material	An optional parameter used to specify the wall material of the target. Any material from the thermal database used to represent wall materials may be used here. Since the transient heat conduction problem is not solved now for the target this parameter is not used.
Method	indicates the solution method STEADY for steady state XPLICIT for explicit (as is done with species) MPLICIT for implicit (as is done with all other variables) Default method is MPLICIT

148

Equation Type	If METHOD is not STEADY, this parameter further indicates the equation type to be either ODE or PDE. The default is PDE

Key word: TARG	
Input: Compartment, Surface, Depth Position, Breadth Position, Method, Equation Type	
Compartment	The compartment in which the target is located
Surface	Surface defines the compartment surface for the target. It is one of the character strings, UP, FRONT, RIGHT, BACK, LEFT and DOWN
Depth Position	Position of the target as a distance from the rear wall of the target compartment (X direction). Default value is ½ compartment depth.
Breadth Position	Position of the target as a distance from the left wall of the target compartment (Y direction). Default value is ½ compartment breadth.

Method	indicates the solution method
	     STEADY for steady state
	     XPLICIT for explicit (as is done with species)
	     MPLICIT for implicit (as is done with all other variables)
	
	Default method is MPLICIT

Equation Type	If METHOD is not STEADY, this parameter further indicates the equation type to be either ODE or PDE. The default is PDE

B.16 Modeling Compartment as a Shaft or Hallway

For stairwells, elevator shafts, and similar compartments, the use of a single, well-mixed zone better approximates conditions within the compartment. To specify use of a one zone model for individual compartments rather than the typical two zone model, the SHAFT keyword is used.

Example:

SHAFT 1

149

Key word: SHAFT Input: Compartment	
Compartment	Compartment to be modeled as a single, well-mixed zone

For long hallways or corridors, there can be a significant delay time for the initial hot gas layer to travel along the ceiling to the far end of the compartment. To allow the CFAST model to calculate the ceiling jet velocity and temperature along the corridor and its effects on heat transfer to the ceiling, the HALL keyword is used. In addition, the horizontal position of the vents in the compartment are also specified with the HVENT keyword.

Example:

HALL 1

Key word: HALL Input: Compartment, Velocity, Depth, Decay Distance	
Compartment	Compartment to be modeled as a single, well-mixed zone
Velocity	Optional parameter, ceiling jet velocity at the distance from the reference point where the temperature falls off by 50 %.
Depth	Optional parameter, ceiling jet depth at the distance from the reference point where the temperature falls off by 50 %.
Decay Distance	Optional parameter, distance from the reference point where the temperature falls off by 50 %.

B.17 Runtime Graphics

A graphics specification can be added to the data file. Details of the meaning of some of the parameters is best left to the discussion of the device independent graphics software used by CFAST. However, the information necessary to use it is straightforward. The general structure is similar to that used for the compartment and fire specification. One must tell the program "what to plot," "how it should appear," and "where to put it."

The key words for "where to put it" are:

150

DEVICE	where to plot it
BAR	bar charts
GRAPH	specify an x-y plot
TABLE	put the data into a table
PALETTE	specify the legend for CAD views
VIEW	show a perspective picture of the structure
WINDOW	the size of the window in "user" space.

The complete key word is required. That is, for the "where to put it" terms, no abbreviations are allowed. Then one must specify the variables to be plotted. They are:

VENT, HEAT, PRESSUR, WALL, TEMPERA, INTERFA,
H_2O, CO_2, CO, OD, O_2, TUHC, HCN, HCL, CT

As might be expected, these are similar key words to those used in the plotting program, CPlot. In this case, it is a reduced set. The application and use of CFAST and CPlot are different.

For each key word there are parameters to specify the location of the graph, the colors and finally, titles as appropriate. For the variables, there is a corresponding pointer to the graph of interest.

The WINDOW label specifies the user space for placement of graphs, views, *etc.* The most common values (which are also the default) are:

$Xl = 0.$, $Yb = 0.$, $Zf = 0.$
$Xr = 1279.$, $Yt = 1023.$, $Zb = 10.$

This is not a required parameter; however, it is often convenient to define graphs in terms of the units that are used. For example, if one wished to display a house in terms of a blueprint, the more natural units might be feet. In that case, the parameters might have the values:

$Xl = 0.$, $Yb = 0.$, $Zf = 0.$
$Xr = 50.$, $Yt = 25.$, $Zb = 30.$

Up to five graphs, tables, bar charts, and views may be displayed at one time on the graphics display. Up to five labels may be displayed at one time on the graphics display. Each type of output and each label is identified by a unique number (1-5) and placed in the window at a specified location. Xl, Yb, Zf, Xr, Yt and Zb have a meaning similar to WINDOW. However, here they specify where in the window to put the output.

The PALETTE label performs a specialized function for showing colors on the views. A four entry table is created and used for each type of filling polygon used in a view. Up to five palettes may be

151

defined. Each palette is identified by a unique number and placed in the window at a specified location. Xl, Yb, Zf, Xr, Yt and Zb have a meaning similar to WINDOW. However, here they specify where in the window to put the palette.

In order to see the variables, they must be assigned to one of the above displays. This is accomplished with the variable pointers as:

```
(Variable) (nmopq) (Compartment) (Layer).
```

Variable is one of the available variables VENT, HEAT, PRESSUR, WALL, TEMPERA, INTERFA, N_2, O_2, CO_2, CO, HCN, HCL, TUHC, H_2O, OD, CT used as a label for the line. The species listed correspond to the variable "SPECIES" in CPlot. (nmopqr) is a vector which points to:

```
index     display in

(1) n -> bar chart
(2) m -> table
(3) o -> view
(4) p -> label
(5) q -> graph
```

respectively. These numbers vary from 1 to 5 and correspond to the value of "n" in the "where to put it" specification. Compartment is the compartment number of the variable and Layer is "U" or "L" for upper and lower layer, respectively.

Example:

```
WINDOW          0      0  -100  1280  1024   1100
GRAPH 1  100.  050.  0.   600.  475.   10.   3 TIME   HEIGHT
GRAPH 2  100.  550.  0.   600.  940.   10.   3 TIME   CELSIUS
GRAPH 3  720.  050.  0.  1250.  475.   10.   3 TIME   FIRE_SIZE(kW)
GRAPH 4  720.  550.  0.  1250.  940.   10.   3 TIME   O|D2|O(%)
INTERFA 0 0 0 0 1    1 U
TEMPERA 0 0 0 0 2    1 U
HEAT    0 0 0 0 3    1 U
O2      0 0 0 0 4    1 U
INTERFA 0 0 0 0 1    2 U
TEMPERA 0 0 0 0 2    2 U
HEAT    0 0 0 0 3    2 U
O2      0 0 0 0 4    2 U
```

Key word: DEVICE Input: Plotting Device	
Plotting Device	The Plotting Device specifies the hardware device where the graphics is to be displayed. It is installation dependent. In general it specifies which device will receive the output. For most systems, 1 is for the screen from which keyboard input comes, and 6 is for the hpgl files.

Key word: WINDOW Inputs: Xl, Yb, Zf, Xr, Yt, Zb	
Xl	Left hand side of the window in any user desired units.
Yb	Bottom of the window in any user desired units.
Zf	Forward edge of the 3D block in any user desired units.
Xr	Right hand side of the window in any user desired units.
Yt	Top of the window in any user desired units.
Zb	Rear edge of the 3D block in any user desired units. These definitions refer to the 3D plotting block that can be seen.

Key word: BAR	
Inputs: Bar Chart Number, Xl, Yb, Zf, Xr, Yt, Zb, Abscissa Title, Ordinate Title	
Bar Chart Number	The number to identify the bar chart. Allowable values are from 1 to 5.
Xl	Left hand side of the bar chart within the window in the same units as that of the window.
Yb	Bottom of the bar chart within the window in the same units as that of the window.
Zf	Forward edge of the 3D block within the window in the same units as that of the window.
Xr	Right hand side of the bar chart within the window in the same units as that of the window.
Yt	Top of the bar chart within the window in the same units as that of the window.
Zb	Back edge of the 3D block within the window in the same units as that of the window.
Abscissa Title	Title for the abscissa (horizontal axis). To have blanks in the title, use the underscore character " _ ".
Ordinate Title	Title for the ordinate (vertical axis). To have blanks in the title, use the underscore character "_".

Key word: GRAPH	
Inputs: Graph Number, Xl, Yb, Zf, Xr, Yt, Zb, Color, Abscissa Title, Ordinate Title	
Graph Number	The number to identify the graph. Allowable values are from 1 to 5. The graphs must be numbered consecutively, although they do not have to be given in order. It is acceptable to define graph 4 before graph 2, but if graph 4 is to be used, then graphs 1 through 3 must also be defined.
Xl	Left hand side of the graph within the window in the same units as that of the window.
Yb	Bottom of the graph within the window in the same units as that of the window.
Zf	Forward edge of the 3D (three dimensional) block within the window in the same units as that of the window.
Xr	Right hand side of the graph within the window in the same units as that of the window.
Yt	Top of the graph within the window in the same units as that of the window.
Zf	Back edge of the 3D block within the window in the same units as that of the window.
Color	The color of the graph and labels which is specified as an integer from 1 to 15. Refer to DEVICE (NBSIR 85-3235) for the colors corresponding to the color values.
Abscissa Title	Title for the abscissa (horizontal axis). To have blanks in the title, use the underscore character " ".
Ordinate Title	Title for the ordinate (vertical axis). To have blanks in the title, use the underscore character "_".

155

Key word: TABLE	
Inputs: Table Number, Xl, Yb, Zf, Xr, Yt, Zb	
Table Number	The table number is the number to identify the table. Allowable values are from 1 to 5. The tables must be numbered consecutively, although they do not have to be given in order. It is acceptable to define table 4 before table 2, but if table 4 is to be used, then tables 1 through 3 must also be defined.
Xl	Left hand side of the table within the window in the same units as that of the window.
Yb	Bottom of the table within the window in the same units as that of the window.
Zf	Forward edge of the 3D block within the window in the same units as that of the window.
Xr	Right hand side of the table within the window in the same units as that of the window.
Yt	Top of the table within the window in the same units as that of the window.
Zb	Back edge of the 3D block within the window in the same units as that of the window.

Key word: VIEW Inputs: View Number, Xl, Yb, Zf, Xr, Yt, Zb, File, Transform Matrix	
View Number	View number is the number to identify the view. Allowable values are from 1 to 5. The views must be numbered consecutively, although they do not have to be given in order. It is acceptable to define view 4 before view 2, but if view 4 is to be used, then views 1 through 3 must also be defined.
Xl	Left hand side of the view within the window in the same units as that of the window.
Yb	Bottom of the view within the window in the same units as that of the window.
Zf	Forward edge of the 3D block within the window in the same units as that of the window.
Xr	Right hand side of the view within the window in the same units as that of the window.
Yt	Top of the view within the window in the same units as that of the window.
Zf	Back edge of the 3D block within the window in the same units as that of the window.
File	File is the filename of a building descriptor file.
Transform Matrix	The Transform Matrix is a 16 number matrix which allows dynamic positioning of the view within the window. The matrix (1 0 0 0 0 1 0 0 0 0 1 0 0 0 0 1) would show the image as it would appear in a display from BUILD.

Key word: LABEL Inputs: Label Number, Xl, Yb, Zf, Xr, Yt, Zb, Text, Angle1, Angle2	
Label Number	Label number is the number to identify the label. Allowable values are from 1 to 5.
Xl	Left hand side of the label within the window in the same units as that of the window.
Yb	Bottom of the label within the window in the same units as that of the window.
Zf	Forward edge of the 3D block within the window in the same units as that of the window.
Xr	Right hand side of the label within the window in the same units as that of the window.
Yt	Top of the label within the window in the same units as that of the window.
Zb	Back edge of the 3D block within the window in the same units as that of the window.
Text	The text to be displayed within the label. To have blanks in the title, use the underscore character " ".
Color	Color of the text to be displayed (a number from 0 to 15).
Angle1 and Angle2	Angles for display of the label in a right cylindrical coordinate space. At present, only the first angle is used and represents a positive counter-clockwise rotation; set the second angle to zero. Both angles are in radians.

Key word: PALETTE **Inputs:** Palette Number, Xl, Yb, Zf, Xr, Yt, Zb, Color and Label	
Palette Number	Palette number is the number to identify the palette. Allowable values are from 1 to 5.
Xl	Left hand side of the palette within the window in the same units as that of the window.
Yb	Bottom of the palette within the window in the same units as that of the window.
Zf	Forward edge of the 3D block within the window in the same units as that of the window.
Xr	Right hand side of the palette within the window in the same units as that of the window.
Yt	Top of the palette within the window in the same units as that of the window.
Zb	Back edge of the 3D block within the window in the same units as that of the window.
Color and Label	There are four pairs of color/text combinations, each corresponding to an entry in the palette. The color number is an integer from 1 to 15 and the text can be up to 50 characters (total line length of 128 characters maximum). As before, spaces are indicated with an underscore character " ".

Appendix C CFAST Thermal Properties Database Format

Thermal data is read from a file which is in an ASCII format. The default name is THERMAL.DF. An alternative can be chosen by using the key word THRMF in the data file that is used by the CFAST model.

The relationship is by the name used in specifying the boundary. Any name can be used so long as it is in the thermal database. If a name is used which is not in the database, then FAST will turn off the conduction calculation, and FAST will stop with an appropriate error message. The form of an entry in the database is:

name conductivity specific heat density thickness emissivity

and the units are:

name	1 to 8 alphanumeric characters
conductivity	Watts/meter/Kelvin
specific heat	Joules/kilogram/Kelvin
density	kilograms/cubic meter
thickness	meters
emissivity	dimensionless.

Appendix D CFAST Object Fire Database Format

The object database was modified to handle the new definition of a flame spread object. Please note that the format for the objects database for CFAST 2.2 (and earlier) and 3.0 are not compatible.

line 1) OBJECT OBJNAM

      OBJNAM(NUMOBJL) = Name of object (up to 8 characters)

line 2) OBJTYP OBJCRI(2,I) OBJCRI(3,I) OBJMAS OBJGMW OBJVT OBJHC

      OBJTYP(NUMOBJL) = Object type
          1 unconstrained burn
          2 constrained burn
          3 flame spread model
      OBJCRI(2,NUMOBJL) = Flux for ignition (W/m²)
      OBJCRI(3,NUMOBJL) = Surface temperature for ignition (k)
      OBJMAS(NUMOBJL)   = Total mass (kg)
      OBJGMW(NUMOBJL)   = Gram molecular weight
      OBJVT(NUMOBJL)   = Volatilization temperature (k)
      OBJHC(NUMOBJL)   = Heat of combustion (J/kg)

line 3) OBJXYZ(1,I) OBJXYZ(2,I) OBJXYZ(3,I) OBJORT(1,I) OBJORT(2,I) OBJORT(3,I)

      OBJXYZ(1,NUMOBJL) = Panel length (m)
      OBJXYZ(2,NUMOBJL) = Panel height or width (m)
      OBJXYZ(3,NUMOBJL) = Panel thickness (m)
      OBJORT(1,NUMOBJL) = x component of the normal
      OBJORT(2,NUMOBJL) = y component of the normal
      OBJORT(3,NUMOBJL) = z component of the normal

For type 1 and 2 fires line 4) is blank. For type 3 objects line 4 has data specific only to flame spread. The format for a flame spread object follows.

line 4) OBJMINT OBJPHI OBJHGAS OBJQAREA

      OBJMINT = Minimum surface temperature for lateral flame spread (K)
      OBJPHI = Lateral flame spread parameter (J^2/s^2*m^3)
      OBJHGAS = Effective heat of gasification (J/kg)

163

OBJQAREA   = Total heat per unit area (J/m^2)

The effective heat of gasification is the one unusual parameter for the flame spread object. The method given by Quintiere and Cleary is to plot the peak heat release rate (HRR) against the radiant flux used for a number of cone calorimeter runs at different fluxes. The slope of the curve that bests fits the data is the effective heat of gasification.

line 5) OTIME(1,I) to OTIME(TOTJ,I)

OTIME(NV,NUMOBJL) = Time history (s)

Each OTIME(J,I) represents a point on the objects burn time line where the variables below are defined exactly. CFAST will interpolate values between any two points. TOTJ is the total number of points on the specified time line. CFAST automatically assigns an initial time zero for the objects time line so that there will always be one fewer specified value for the time line than for the history variables below. For type three objects the curves are straight lines to represent the constants that CFAST uses for the flame spread objects.

The following lines are the histories for the individual parameters at each of the OTIME points.

Line 6) OMASS(1,I) to OMASS(TOTJ+1,I)

OMASS(NV,NUMOBJL) = Pyrolysis rate time history (kg/s)

line 7) OQDOT(1,I) to OQDOT(TOTJ,I)

OQDOT(NV,NUMOBJL) = Rate of heat release time history (w)

line 8) OAREA(1,I) to OAREA(TOTJ,I)

OAREA(NV,NUMOBJL) = Area of fire time history (m^2)

line 9) OHIGH(1,I) to OHIGH(TOTJ,I)

OHIGH(NV,NUMOBJL) = Height of flame time history (m)

line 10) OCO(1,NUMOBJL) to OCO(TOTJ,NUMOBJL)

OCO(NV,NUMOBJL) = CO/CO2 time history

line 11) OOD(1,NUMOBJL) to OOD(TOTJ,NUMOBJL)

OOD(NV,NUMOBJL) = OD or soot time history

line 12) OHRC(1,NUMOBJL) to OHRC(TOTJ,NUMOBJL)

OHCR(NV,NUMOBJL) = H/C time history

line 13) OOC(1,I) to OOC(TOTJ,I)

OOC(NV,NUMOBJL) = O/C time history

line 14) OMPRODR(1,10,I) to OMPRODR(TOTJ,10,I)

OMPRODR(NV,10,NUMOBJL) = CT time history

line 15) OMPRODR(1,5,I) to OMPRODR(TOTJ,5,I)

OMPRODR(NV,5,NUMOBJL) = HCN time history

line 16) OMPRODR(1,6,I) to OMPRODR(TOTJ,6,I)

OMPRODR(NV,6,NUMOBJL) = HCl time history

Appendix E Alphabetical Listing of CFAST Routines

NAME	LOCATION	DESCRIPTION
ATMOSP	ATMOSP.SOR	Floating point function to calculate the pressure, temperature and density at a specified height.
AVG	NAILED.SOR	Calculate the average of two real numbers.
CEILHT	CEILHT.SOR	Calculate convective heat transfer to the uniform temperature ceiling above a fire in a parallelepiped compartment with a two-layer fire environment.
CFAST	CFAST.SOR	Main model.
CHEMIE	CHEMIE.SOR	Do the combustion chemistry - for plumes in both the upper and lower layers.
CJET	CJET.SOR	Interface between RESID and CEILHT; sets up variables to pass.
CNDUCT	CNDUCT.SOR	Heat conduction through objects.
CNHEAT	CNHEAT.SOR	Interface routine between RESID and CONVEC, CNDUCT for surfaces.
CONHT	NAILED.SOR	Calculate fractional effective dose due to temperature in 1 minute in REPORT.
CONT1	NAILED.SOR	Obtain data for new time step in REPORT.
CONVEC	CONVEC.SOR	Convective heat loss or gain.
CONVRT	CONVRT.SOR	Convert an ASCII string to an integer or floating point number.
COPY	NAILED.SOR	Copy one matrix to another in REPORT.
CPTIME	CPTIME.SOR	Return the total calculation (not simulation) time since the beginning.

NAME	LOCATION	DESCRIPTION
CUNITS	CUNITS.SOR	Convert from user units to scientific units or vice versa.
CVHEAT	CVHEAT.SOR	Interface between RESID and CONVEC; sets up variables to pass.
D1MACH	AAUX.SOR	Used to obtain machine-dependent parameters for the local machine environment.
DATACOPY	DATACOPY.SOR	Copy the solver variables to the environment common blocks.
DATYPE	DATYPE.SOR	Determine the type of the ASCII string (integer,floating point,...).
DDASSL	DASSL.SOR	Solver.
DEBUGPR	SOLVE.SOR	Diagnostic routine for responding to function keys in CFAST.
DEFAULT	DEFAULT.SOR	Allow the user to set CPlot default parameters for the compartment number, layer, species, vent flow destination, and character set.
DELP	DELP.SOR	Calculate the absolute hydrostatic pressures at a specified elevation in each of two adjacent compartments and the pressure difference.
DISCLAIM	DISCLAIM.SOR	Disclaimer notice.
DJET	DJET.SOR	Physical interface routine to calculate the current rates of mass and energy flows into the layers from all door jet fires in the structure.
DJFIRE	DJFIRE.SOR	Calculate heat and combustion chemistry for a door jet fire.
DOFIRE	DOFIRE.SOR	Do heat release from a fire for both main fire and objects.
DOSES	NAILED.SOR	Determine hazard the person is exposed to based on layer; used in REPORT.
DREADIN	DREADIN.SOR	Read a record (binary) of a history file. See WRITEOT.
DUMPER	DUMPER.SOR	Write to the history file.

NAME	LOCATION	DESCRIPTION
ENTRAIN	ENTRAIN.SOR	Calculate entrainment for HFLOW.
ENTRFL	ENTRFL.SOR	Low-level routine with ENTRAIN to calculate vent entrainment.
FDCO	NAILED.SOR	Fractional effective dose due to CO in 1 minute in REPORT.
FDCO2	NAILED.SOR	Fractional effective dose due to CO2 in 1 minute in REPORT.
FDHCN	NAILED.SOR	Fractional effective dose due to HCN in 1 minute in REPORT.
FDO2	NAILED.SOR	Fractional effective dose due to O2 in 1 minute in REPORT.
FIND	NAILED.SOR	Verify a particular type of keyword in REPORT; used by MATCH.
FIRES	FIRES.SOR	Physical interface routine to calculate the current rates of mass and energy flows into the layers from all fires in the building.
FIRPLM	FIRPLM.SOR	Calculates plume entrainment for a fire from McCaffrey's correlation.
FLAMHGT	FLAMHGT.SOR	Calculates the flame height for a given fire size and area.
FLOGO1	FLOGO1.SOR	Deposition of mass, enthalpy, oxygen, and other product-of-combustion flows passing between two compartments through a vertical, constant-width vent.
FLWOUT	FLWOUT.SOR	Display the flow field in CFAST.
GASLOAD	GASLOAD.SOR	Read species keyword specifications in graphics section of input file; used by NPUTG
GETELEV	GETELEV.SOR	Set elevation coordinates for horizontal flow vent; called by VENT.
GETOBJ	GETOBJ.SOR	Get database detailed information for a specified object.

NAME	LOCATION	DESCRIPTION
GETVAR	GETVAR.SOR	Calculate conditions for specified compartment; used by HFLOW.
GJAC	GJAC.SOR	Evaluate the Jacobian for the solver. (Not used at this time).
GRAFIT	GRAFIT.SOR	Initialize a graphics grid or contour diagram with a full set of labels and switch the rest of the contour graphics material to the graphics mode; modified version for CFAST of GRAFIT from DEVICE.
GRES	GRES.SOR	User-supplied subroutine to calculate the functions; SNSQE routine.
GRES2	GRES2.SOR	User-supplied subroutine to calculate the functions; SNSQE routine.
GTOBST	GTOBST.SOR	Read a line from the objects database; used with GETOBJ.
HALLHT	HALLHT.SOR	Calculates the velocity and temperature of the ceiling jet at each detector location in a corridor.
HCL	HCL.SOR	Physical interface routine to do HCl deposition on wall surfaces.
HCLTRAN	HCLTRAN.SOR	Calculate the hydrogen chloride balance in the gas and on the wall surface; called by HCL.
HFLOW	HFLOW.SOR	Physical interface routine to calculate flow through all unforced horizontal vents.
HVFAN	HVFAN.SOR	Calculate mass flow rate through a fan based on a fan curve.
HVFREX	HVFREX.SOR	Update arrays and assign compartment pressures, temperatures and species to HVAC exterior nodes (from compartments)
HVFRIC	HVFRIC.SOR	Calculate duct friction factor from the Colebrook equation.
HVINIT	HVINIT.SOR	Set up the arrays needed for HVAC simulation and initialize temperatures and concentrations.

NAME	LOCATION	DESCRIPTION
HVMAP	HVMAP.SOR	Construct mapping arrays used to map interior nodes, exterior nodes, and HVAC systems to appropriate arrays.
HVMFLO	HVMFLO.SOR	Compute the source term for mass flow in the mechanical ventilation system.
HVSFLO	HVSFLO.SOR	Compute the source term for temperature in the mechanical ventilation system.
HVTOEX	HVTOEX.SOR	Update arrays and assign compartment pressures, temperatures and species to HVAC exterior nodes (to compartments)
HVVIS	HVVIS.SOR	Calculate viscosity of air; used by HVMFLO.
INDEXI	INDEXI.SOR	Create the index array used by SORTFR.
INITAMB	INITAMB.SOR	Initialize the ambient conditions based on TAMB and EAMB.
INITOB	INITOB.SOR	Initialize object data.
INITMM	INITMM.SOR	Initialize main memory; required for all modules which will run the CFAST kernel.
INITSLV	INITSLV.SOR	Initialization for solver variables and arrays.
INITSOLN	INITSOLN.SOR	Determine an initial solution to the zone fire modeling equations.
INITSPEC	INITSPEC.SOR	Initialize variables associated with species, based on the ambient conditions.
INITTARG	INITTARG.SOR	Initialize target data.
INITWALL	INITWALL.SOR	Initialize wall variables and arrays, based on conduction settings.
INT2D	CEILHT.SOR	Integrate a function over a region formed by intersecting a rectangle and a circle; used by CEILHT.
INTERP	INTERP.SOR	Interpolate a table of numbers found in two arrays.
INTSQ	CEILHT.SOR	CEILHT integration routine; used by INT2D.
INTTABL	CEILHT.SOR	CEILHT integration routine; used by INT2D.

NAME	LOCATION	DESCRIPTION
INTTRI	CEILHT.SOR	CEILHT integration routine; used by INT2D.
JAC	JAC.SOR	Evaluate the Jacobian for the solver. (Not used at this time).
KILLPACK	WRITEOT.SOR	Display "FATAL ERROR" message when writing compressed history record in CFAST.
LENGTH	LENGTH.SOR	Calculate length of character string.
LENOCO	LENOCO.SOR	Function to calculate length of the main common block (MOCO1A) in storage units.
LINTERP	ABSORB.SOR	Linearly interpolate absorption data.
LOADUP	LOADUP.SOR	Load numeric values from graphics descriptor lines into appropriate arrays; used by NPUTG.
MAKTABL	CEILHT.SOR	Set up the integration table used by INTTABL in CEILHT.
MESS	MESS.SOR	Write a string to IOFILO.
MVENT	MVENT.SOR	Physical interface routine for the HVAC model.
MVOLAST	MVOUT.SOR	Display mechanical ventilation elevation information.
MVOUT	MVOUT.SOR	Display the mechanical ventilation information from NPUTO in CFAST.
NPUTG	NPUTG.SOR	Read and sort out the graphics descriptors in the input file for display routines in CFAST.
NPUTO	NPUTO.SOR	Write to IOFILO a summary of information read from input file in CFAST.
NPUTOB	NPUTOB.SOR	Read the objects database and initialize arrays for selected objects only in CFAST.
NPUTP	NPUTP.SOR	Control reading of input file with a post-read verification and cleanup.
NPUTQ	NPUTQ.SOR	Read all input file information except the graphics descriptors.
NPUTT	NPUTT.SOR	Read the complete thermophysical properties database into memory for verification.

NAME	LOCATION	DESCRIPTION
NTRACT	SOLVE.SOR	Key pressed in SOLVE, determine what to do.
OBJDFLT	OBJDFLT.SOR	Set up values for "DEFAULT" entry in objects database.
OBJFND	OBJFND.SOR	Search objects database for object name and return pointer to record number.
OBJINT	OBJINT.SOR	Interpolating routine for object fire time history values. Similar to PYROLS.
OBJOUT	OBJOUT.SOR	Write to IOFILO a summary of the objects information from database for specified objects.
OBJSHFL	DISOB1.SOR	Shuffle the ODBNAM and ODBREC arrays if objects are deleted or invalidated.
OBTYPE	OBTYPE.SOR	Determine the type of a specified object.
OFFSET	OFFSET.SOR	Calculate the solver array offsets for each variable.
OPENOBJ	OPENOBJ.SOR	Open and read the object database into memory arrays.
OPENSHEL	OPENSHEL.SOR	Read the configuration file, set up the environment, open the output file.
OPUT	WRITEOT.SOR	Set up output arrays in PACKOT for ASCII history file.
OUTAMB	NPUTO.SOR	Output ambient conditions.
OUTCOMP	NPUTO.SOR	Output geometry.
OUTFIRE	NPUTO.SOR	Output initial main fire specifications.
OUTOBJ	NPUTO.SOR	Output object fire specifications.
OUTPU1	OUTPU1.SOR	Obtains a single piece of data for CFAST display.
OUTTHE	NPUTO.SOR	Output thermal specifications.
OUTVENT	NPUTO.SOR	Output vent connections.
PACKOT	WRITEOT.SOR	Compression routine for writing the history file.

NAME	LOCATION	DESCRIPTION
PYROLS	PYROLS.SOR	Interpolating routine for specified fire time history values.
QFCLG	CEILHT.SOR	Compute the convective heat transfer flux to the ceiling at location $(X,Y)=(Z(1),Z(2))$.
RAD2	RAD2.SOR	Compute the radiative heat flux for 2 surfaces (extended ceiling and extended floor). Compute the heat absorbed by the lower and upper layers.
RAD4	RAD4.SOR	Compute the radiative heat flux for 4 surfaces (ceiling, upper wall, lower wall, and floor). Compute the heat absorbed by the layers.
RDABS	RDABS.SOR	Compute the energy absorbed by the upper and lower layer due to radiation given off by heat emitting rectangles forming the enclosure. Called by RAD2 and RAD4.
RDFANG	RDFIGSOL.SOR	Define solid angles for fires; called by RAD4.
RDFLUX	RDFLUX.SOR	Calculate the "C" vector in the net radiation equations of Seigel and Howell and the heat absorbed by the lower and upper layer fires due to gas layer emission and fires.
RDFTRAN	RDTRAN.SOR	Define transmission factors for fires; called by RAD4.
RDHEAT	RDHEAT.SOR	Radiation transfer routine between RESID and RAD2 or RAD4.
RDPARFIG	RDFIGSOL.SOR	Calculate the configuration factor between two parallel plates a distance z apart.
RDPRPFIG	RDFIGSOL.SOR	Calculate the configuration factor between two perpendicular plates with a common edge.
RDRTRAN	RDTRAN.SOR	Define upper layer transmission factors; called by RAD4.
RDSANG	RDFIGSOL.SOR	Used by RDFANG to define solid angles for fires.
RDSANG1	RDFIGSOL.SOR	Used by RDFANG to define solid angles for fires.
READAS	READAS.SOR	Read from IOFILI a single line in ASCII format.

174

NAME	LOCATION	DESCRIPTION
READASTU	READASTU.SOR	Similar to READAS except this contains automatic conversion to upper case. This is to filter commands from the console so they are not case-sensitive.
READBF	READIN.SOR	Determine if string buffer has been fully parsed, if so, read a new ASCII line.
READCF	READCF.SOR	Read the configuration file for OPENSHEL.
READCV	READCV.SOR	Convert data in graphics descriptor lines to floating point or integer format.
READFL	READIN.SOR	Read data from string buffer as an ASCII character string.
READHI	READHI.SOR	Read the history file for specified time step.
READIN	READIN.SOR	Parse the line from READAS.
READOP	READOP.SOR	Read command line options, set up environment, and open designated files.
READRS	READIN.SOR	Restart START value for string parsing.
RESID	RESID.SOR	Calculate the right hand side of the predictive equations.
RESTRT	RESTRT.SOR	Called by CFAST to read to specified interval from a history file when RESTRT is specified.
RESULT	RESULT.SOR	Print out the model results of the simulation at the current time to IOFILO.
RESYNC	RESYNC.SOR	Resynchronize the total mass of the species with that of the total mass.
SETP0	SETP0.SOR	Reset the solver array in the middle of a simulation.
SLVHELP	SOLVE.SOR	Display function key options available within CFAST run.
SNSQE	SNSQE.SOR	Find a zero of a system of N nonlinear functions in N variables by a modification of the Powell hybrid method.

NAME	LOCATION	DESCRIPTION
SOLVE	SOLVE.SOR	Top level of the solver for the predictive equations; Calls RESID, DASSL, *etc.*
SORTFR	SORTFR.SOR	Sort the room and fire related arrays setup in FIRES by increasing room number order.
SQFWST	CEILHT.SOR	Calculate average heat transfer fluxes to lower and upper walls along a vertical line passing through a wall/ceiling jet stagnation point.
SSTRNG	SSTRNG.SOR	Parse a string for space or comma delimited substrings.
SSFIR	SSOUT.SOR	Output the current fire environment to a spreadsheet file.
SSFLW	SSOUT.SOR	Output the current vent flow to a spreadsheet file.
SSLAY	SSOUT.SOR	Output the current two layer environment to a spreadsheet file.
SSOUT	SSOUT.SOR	Output the results of the simulation at the current time to a spreadsheet file.
SSPRO	SSOUT.SOR	Output the current wall temperatures to a spreadsheet file.
SSSP	SSOUT.SOR	Output the current species to a spreadsheet file.
SSSPRINK	SSOUT.SOR	Output the current sprinkler environment to a spreadsheet file.
SSTAR	SSOUT.SOR	Output the current surface target temperatures and fluxes to a spreadsheet file.
TOXIC	TOXIC.SOR	Calculate the parts per million of gases and the concentration time dose.
TOXICB	TOXICB.SOR	Setup and display the labels for table of doses, *etc.* in CFAST.
TOXICH	TOXICH.SOR	Setup and display the table for doses, *etc.* in CFAST.
TOXICR	TOXICR.SOR	Setup and display the values for table of doses, *etc.* in CFAST.
UNPACK	DREADIN.SOR	Decompression routine for reading the history file.

NAME	LOCATION	DESCRIPTION
VDCO2	NAILED.SOR	Multiplication factor for the effect on ventilation allowing for the effect of the increased RMV caused by carbon dioxide on the rate of uptake of other toxic gases.
VENT	VENT.SOR	Calculation of the flow of mass, enthalpy, oxygen and other products of combustion through a vertical, constant-width vent in a wall segment common to two compartments.
VENTCF	VENTCF.SOR	Calculate the flow of mass, enthalpy, and products of combustion through a horizontal flow vent joining an upper space 1 to a lower space 2.
VFLOW	VFLOW.SOR	Interface between CFAST and the vertical flow vent physical routines.
WRITEOT	WRITEOT.SOR	Write a record (binary) of a history file. See DREADIN.
WSET	WSET.SOR	Set Initial wall variables and arrays for one surface; used by INITWALL.
XERRWV	AAUX.SOR	Process an error (diagnostic) message
XFDCO	NAILED.SOR	Fractional effective dose due to CO for specified interval in REPORT; see FDCO.
XFDHCN	NAILED.SOR	Fractional effective dose due to HCN in specified interval in REPORT; see FDHCN.
XFDO2	NAILED.SOR	Fractional effective dose due to O2 in specified interval in REPORT; see FDO2.

Index

179

www.ingramcontent.com/pod-product-compliance
Lightning Source LLC
Chambersburg PA
CBHW081612200526
45167CB00019B/2403